Measuring Welfare beyond Economics

T0330805

Dissatisfaction with the Gross Domestic Product (GDP) as an indicator of a country's development or a population's well-being led to the development of the Genuine Progress Indicator (GPI). The GPI is an aggregate index of over 20 economic, social and environmental indicators, and accounts for both the welfare benefits of economic growth, and the social and environmental costs which accompany that economic growth. The result is better information about the level of welfare or well-being of a country's population.

This book measures the GPI of Hong Kong and Singapore from 1968 to 2010. It finds that for both countries, economic output (as measured by the GDP) has grown more than welfare (as measured by the GPI), but important differences are also found. In Hong Kong, the GPI has grown for the whole period under consideration, while in Singapore the GPI has stalled from 1993. This is in line with most countries and is explained by the 'threshold hypothesis' which states that beyond a certain level of economic development the benefits of further economic growth are outweighed by even higher environmental and social costs. The book argues that the growth of Hong Kong's GPI is due to its favourable relationship with China and in particular its ability to export low-wage jobs and polluting industries, rather than successful domestic policies. A stalling or shrinking GPI calls for alternative policies than the growth economy promoted by neoclassical economists, and the book explores an alternative model, that of the Steady State Economy (SSE).

Claudio O. Delang is Assistant Professor at the Department of Geography of Hong Kong Baptist University.

Yi Hang Yu is Researcher at the Department of Geography of Hong Kong Baptist University.

Routledge Studies in Sustainable Development

This series uniquely brings together original and cutting-edge research on sustainable development. The books in this series tackle difficult and important issues in sustainable development including: values and ethics; sustainability in higher education; climate compatible development; resilience; capitalism and de-growth; sustainable urban development; gender and participation; and well-being.

Drawing on a wide range of disciplines, the series promotes interdisciplinary research for an international readership. The series was recommended in the *Guardian*'s suggested reads on development and the environment.

Institutional and Social Innovation for Sustainable Urban Development
Edited by Harald A. Mieg and Klaus Töpfer

The Sustainable University
Progress and prospects
Edited by Stephen Sterling, Larch Maxey and Heather Luna

Sustainable Development in Amazonia
Paradise in the making
Kei Otsuki

Measuring and Evaluating Sustainability
Ethics in sustainability indexes
Sarah E. Fredericks

Values in Sustainable Development
Edited by Jack Appleton

Climate-Resilient Development
Participatory solutions from developing countries
Edited by Astrid Carrapatoso and Edith Kürzinger

Theatre for Women's Participation in Sustainable Development
Beth Osnes

Measuring Welfare beyond Economics

The genuine progress of Hong Kong and Singapore

Claudio O. Delang and Yi Hang Yu

Routledge
Taylor & Francis Group

LONDON AND NEW YORK

First published 2015
by Routledge

2 Park Square, Milton Park, Abingdon, Oxfordshire OX14 4RN
52 Vanderbilt Avenue, New York, NY 10017

Routledge is an imprint of the Taylor & Francis Group, an informa business

First issued in paperback 2019

Copyright © 2015 Claudio O. Delang and Yi Hang Yu

The right of Claudio O. Delang and Yi Hang Yu to be identified as authors of this work has been asserted by him/her in accordance with sections 77 and 78 of the Copyright, Designs and Patents Act 1988.

All rights reserved. No part of this book may be reprinted or reproduced or utilised in any form or by any electronic, mechanical, or other means, now known or hereafter invented, including photocopying and recording, or in any information storage or retrieval system, without permission in writing from the publishers.

Notice:
Product or corporate names may be trademarks or registered trademarks, and are used only for identification and explanation without intent to infringe.

British Library Cataloguing-in-Publication Data
A catalogue record for this book is available from the British Library

Library of Congress Cataloging-in-Publication Data
A catalog record has been requested for this book

ISBN: 978-0-415-81383-9 (hbk)
ISBN: 978-0-367-33285-3 (pbk)

Typeset in Times New Roman
by Saxon Graphics Ltd, Derby

Contents

Figures

Tables

Preface

After decades of steady economic growth, Hong Kong and Singapore have some of the highest GDP per capita in the Asia-Pacific region. Yet, there is also considerable social discontent, as is shown for example by the three-month long pro-democracy protest in Hong Kong in 2014, and the low support for People's Action Party (the lowest since independence) at the 2011 Singaporean general election. Clearly, things are not as rosy as the GDP figures would make us believe.

In this book we look at an alternative measure of development for Hong Kong and Singapore for the 43 years from 1968 to 2010, to assess the extent to which economic growth was accompanied by an improvement in welfare. We do this by using the Genuine Progress Indicator (GPI), which in addition to figures of economic activities also takes into consideration the environmental and social costs that accompany economic growth. Studies have shown that often as the economy grows, the social and environmental costs grow faster than the economic benefits. Beyond a certain point, this results in a loss of welfare. We find that this is also the case of Singapore since the 1990s, but not of Hong Kong.

The book is organized into eight chapters. In Chapter 1 we discuss how the GDP is calculated, and look at its problems, focusing in particular on those addressed by the GPI. This chapter provides an introduction to the book, by reminding the readers that the GDP is not an indicator of welfare, and GDP growth does not necessarily equate with a 'better' life for the people. In Chapter 2 we introduce some alternative indicators to the GDP, dividing them into those that adjust the GDP and those that replace the GDP, some of which have been developed in Hong Kong and Singapore. By so doing, the chapter allows the reader to identify the similarities and differences between the GPI and other indicators, and the advantages of the GPI as an indicator of welfare.

In Chapter 3 we introduce the GPI and discuss its theoretical justifications, by introducing the concept of welfare, wealth, income and capital. We also discuss the strengths and weaknesses of the GPI. One criticism of the GPI is that it is not a uniform indicator, in that different countries include different items, or methods to give an economic value to each item. We argue that this is not only necessary, since not all countries provide the same data (most data used to estimate the GPI are official, nation-wide statistics), but also helpful, since the GPI can be adapted to the unique characteristics of each country. In Chapter 4 we introduce the various items we use to estimate the GPI of Hong Kong and Singapore, and discuss how these items are calculated. For Hong Kong, we include seven economic, eight social and ten environmental items, and for Singapore we include the same seven economic items, but only six social and seven environmental, because of lack of data. We review the methods used to estimate the economic value of each item. While the GPI is not a standardized indicator, as the GDP is, those who have used it have attempted to retain consistency among countries. We also try to use the same indicators used in other studies, and the same methods, to facilitate comparison among countries.

In Chapters 5 and 6 we present the results of Hong Kong and Singapore, respectively, for the 43 years from 1968 to 2010. We show the changes in value of all indicators, and attempt to give a short explanation for the trends we observe. Each chapter concludes with a brief discussion of the overall trend. We find that Hong Kong has been much more successful than Singapore in improving people's welfare, since the GPI has continued growing throughout the period under consideration, albeit at a slower rate than the GDP. On the other hand, Singapore's GPI has not grown since the 1990s, which means that while the GDP has roughly doubled, welfare has hardly increased.

In Chapter 7 we try to explain the differences between Hong Kong's GPI and Singapore's GPI. It has been observed that when countries reach a high level of economic development the environmental and social costs that accompany economic growth outstrip the economic benefits. This has been called the 'threshold hypothesis'. Singapore seems to have reached that threshold in the 1990s. On the other hand, Hong Kong's GPI has continued to grow. We discuss the reasons for such differences between the two city-states.[1] In particular, we argue that Hong Kong has been able to benefit from its favourable location and relation with China.

A flat or dropping GPI calls for alternative policies than those promoted by neoclassical economists who advocate endless economic growth. In Chapter 8 we introduce the concept of a Steady State Economy (SSE). The steady state is an economy that is geared towards a dynamic equilibrium with the ecosystem that supports it, by means of qualitative improvement of existing goods and capital, instead of an increase in the amount of goods produced, which the growth economy depends on. It emphasizes an improvement in the quality of goods and services, rather than an increase in the number of goods produced, a concept which embodies a sustainable use of natural resources and a more equitable distribution of income among citizens. The chapter argues for the introduction of a SSE in Hong Kong and Singapore, discusses its advantages, and introduces the policies necessary to make the transition.

Claudio O. Delang
Yihang Yu
January 2015

Note

1 Hong Kong is of course not a city-state. It is a Special Administrative Region (SAR) that is part of the People's Republic of China. However, it has retained much autonomy in economic, financial, political and social matters, and here we consider it a city-state.

1 Problems with the Gross Domestic Product

Introduction

The Gross Domestic Product (GDP) was developed by Simon Kuznets, an economist at the National Bureau of Economic Research (US), in the 1930s, as an indicator of national economic output. The creation of the GDP was prompted by the need for a standard measure that would be able to quantify the extent of the economic collapse under way, and could be used to devise policies to improve the economy. Following the establishment of international financial institutions, such as the World Bank and the International Monetary Fund at the Bretton Woods Conference in 1944, the GDP was adopted globally as the standard tool to measure the size of a country's economy.

The GDP is calculated as the sum of all final goods and services produced in an economy in a given period of time (Anielski and Soskolne, 2002; Stiglitz, Sen and Fitoussi, 2009). It thus offers an easy method to capture the total consumption of goods and services in a country, and does so in a way that allows for comparisons to be made with the amounts of previous years. The aggregate value can help economists, researchers and the like to assess the level of production across industries, as well as the magnitudes of consumption of a range of goods and services, from baked goods to television units to clinical check-ups. The use of market prices as the unit of measurement also reflects the relative changes in real prices of the different goods and services consumed throughout different time periods (Stiglitz *et al.*, 2009). The GDP, therefore, is an easy to use single number to assess how 'well-off' a society is at a particular moment in time. As a clear and one-dimensional economic indicator, it can also help economists and policy-makers plan the economy and set economic policies to achieve further growth (Hamilton, 1997; Anielski, 2001; Stigliz *et al.*, 2009; Berik and Gaddis, 2011).

For over 70 years the GDP has been used by governments of different countries to assess the success of their monetary and fiscal policies and to draft their national budgets (McCulla and Smith, 2007). International institutions such as the International Monetary Fund (IMF) and the World Bank also use changes in a nation's GDP as an important criterion to fund projects around the world. A recent report by the World Bank stated that high rates of GDP growth are indeed the solution for the world's poverty problem (Commission on Growth and

Development, 2008). Today, the GDP is regularly referred to by politicians, economists, policy-makers and the media as the ultimate metric of a country's welfare and well-being. However, we will argue in this chapter that the GDP is not a suitable indicator, neither of economic development, nor of welfare or well-being, and that it should be abandoned. This chapter brings together and summarizes criticism raised against the use of the GDP, while providing the necessary basis for the discussions on the Genuine Progress Indicator (GPI) that will take place in the following chapters.

The chapter is organized as follows: first, we discuss how the GDP is calculated; second, we discuss the problems with the GDP as an indicator of economic welfare or progress, from a social and environmental perspective. This section addresses the question of why the GPI is a better indicator of welfare than the GDP. In section three we examine the influence that focusing on the GDP has had on economic policies. We also discuss the advantages of abandoning the GDP.

The methodology of the GDP

The GDP accounts for the total value of goods and services produced in an economy during the accounting period. It can be measured in three ways, all of which, in principle, should give the same results: the production (or output) approach, the income approach, and the expenditure approach (Vu, 2009; Geostat, 2011).

The production approach measures GDP as the sum of all 'added value' in the economy – the figure results from subtracting the cost of the goods and services used in production from the value of the total output produced during a particular accounting period (Vu, 2009). For example, if a value of 100 is given to the total output of goods and services in an economy, and the cost of goods and services used in the process of production is 70, then the value added is 30 (Vu, 2009). Taxes on products and import (VAT, excise tax and customs duties) are added, and subsidies on products are subtracted. The equation for this approach is as follows (Geostat, 2011):

> Total GDP at market prices = Total output (goods and services) by types of activities at market prices – intermediary consumption for generating goods and services + taxes on products and import – subsidies on products

The income approach requires information on the factors that are directly involved in the production of goods and services during an accounting period, presented as the sum of all types of factor incomes (returns from resources, or factors of production) generated in the production process, for example 'wages and bonuses and other compensation payable to employees, taxes on products and production payable to the government, and operating surplus for the producers' (Vu, 2009: 5; Geostat, 2011). The equation for this approach is as follows (Geostat, 2011):

Total GDP at market prices = Employment income in the form of wages and Social benefits (including Income tax) + Mixed income received from self-employment + Total profit received by companies from economic activities + Taxes on production and import − Subsidies on production and import

The expenditure approach calculates GDP as the total value of goods and services that are used for final consumption, gross capital formation or total value of transfer of personal savings to business through different types of investments, such as bank deposits, plus net exports (Vu, 2009). The equation of this approach is (Geostat, 2011):

Total GDP at market prices = Consumption expenditure of households + Services rendered by non-profit institutions serving households + Collective and personal services rendered by General Government + Gross capital formation + Changes in inventories + Exports of goods and services − Imports of goods and services

It is important to note that the GDP as a measure of economic activity is not inherently bad. The GDP measures what it was intended to: the size of a country's economic output (the GDP takes into account only monetary transactions of final goods and services produced in an economy). However, it is not (and was never meant to be) a measure of welfare, even though it seems to be treated as such by economists, politicians, and the press. In the next section we discuss its weaknesses, and why it is not a measure of welfare.

Problems of the GDP as an indicator of social welfare

While the GDP is an indicator of economic output, it only includes the market values of traded products. For example, only the costs of extraction of natural resources are included. The inherent values (or existence values (Davidson, 2013)) of these natural resources are not included, nor are their non-marketed qualities (e.g. the cooling properties of urban trees), and non-marketed products (e.g. wild plants consumed by the gatherers). Its creator already acknowledged severe limitations of the GDP: after presenting an itemized list of the things measured by the GDP to Congress in 1934, Simon Kuznets discussed the uses and limitations of the GDP. Kuznets (1934) acknowledged 'a number of other services, in addition to those [itemized goods] listed above might also be considered a proper part of the national economy's end-product'. Kuznets named these services as 'services of housewives and other members of the family', 'relief and charity', 'services of owned durable goods', 'earnings from odd jobs', and 'earnings from illegal pursuits' among others (pp. 3–5). Kuznets cited various reasons for excluding these services from the GDP, the most important of which being his objective of creating an indicator designed to measure only a society's ability to produce and consume goods.

Kuznets thought that the simplicity of a GDP value would make it vulnerable to misrepresentation and detract from its limitations. This was reflected in Kuznets' report to the US Senate, where he stated:

> The valuable capacity of the human mind to simplify a complex situation in a compact characterization becomes dangerous when not controlled in terms of definitely stated criteria. With quantitative measurements especially, the definiteness of the result suggests, often misleadingly, a precision and simplicity in the outlines of the object measured. Measurements of national income are subject to this type of illusion and resulting abuse, especially since they deal with matters that are the center of conflict of opposing social groups where the effectiveness of an argument is often contingent upon oversimplification.
>
> (Kuznets, 1934: 5–6)

In spite of these admonitions, many people consider the GDP a measure of welfare. As such, the focus of economic policies pursued by economists and politicians has for the most part remained set on increasing the growth rates of their country's GDP. Costanza *et al.* (2009) conducted the most comprehensive study on the limitations of the GDP as an indicator of social welfare, and concluded that:

> the GDP ignores changes in the natural, social, and human components of community capital on which the community relies for continued existence and well-being. As a result, GDP not only fails to measure key aspects of the quality of life; in many ways, it encourages activities that are counter to long-term community well-being.
>
> (p. 9)

According to Stockhammer *et al.* (1997), the critical views on the GDP have been exacerbated by the widening gap between economic growth and quality of life since the 1970s. Anielski and Soskolne (2002) argued that well-being is more than economic output and involves multiple causal pathways, which remain unaccounted for in the GDP. These unaccounted pathways include benefits such as unpaid labour, ecosystem services, costs such as crime and environmental degradation, investments in infrastructure, conservation practices and income equality, among others.

In the following pages we describe some of the shortcomings of the GDP as a measure of welfare, dividing them into two different categories: environmental and socioeconomic. These issues, ignored by the GDP, are included in the GPI, as Chapter 4 will elaborate.

Environmental problems with the GDP

1. The GDP acccounts for the flow, but not for stock of resources

The natural resources that are not transformed, and for which no money is invested, are not included in the GDP. For example, the value of timber found in a forest, or the services the forest provides (e.g. in absorbing carbon dioxide) are not included in the GDP, but when that forest is cut and the timber sold, the value of the timber shows up in the GDP. In reality, since a forest has values beyond the commercial value of the timber (for example the value of the carbon sequestered by the vegetation, the forest products that may be used by local populations, soil stabilization, and climate regulations), a country may as well be worse off after the forest is cut, even though the GDP has grown.

2. The GDP encourages the depletion of natural resources

It follows from the previous point that measuring a country's wealth using GDP encourages the depletion of natural resources. When political authorities aim at maximizing GDP, the easiest way to do so is by increasing the extraction and usage of natural resources.

3. The GDP does not account for environmental degradation

By ignoring the stock of natural resources in the GDP, the GDP does not give a clear signal about the conditions of the stock of natural resources in a country. In theory prices should increase as natural resources become scarcer, and therefore more expensive to extract. However, as the technology improves, extraction costs may decrease, which may result in lower prices, even as the natural resources become scarcer. Market prices do not reflect the relative scarcity of natural resources. Hence, one cannot gauge from the GDP whether the environment is being degraded, and the extent and pace at which this is taking place.

4. Environmental degradation increases the GDP

Economic activities place stress on the environment, thereby reducing the ecosystem services that are of value to society. The value of ecosystem services, such as the sequestration of carbon dioxide, the absorption of pollution, the production of oxygen, and the preservation of biodiversity, is very large. Costanza *et al.* (1997) estimated the value of the world's ecosystem services and natural capital at a staggering US$33 trillion per year, larger than the world GDP at that time.

Many natural resources perform particularly valuable non-marketed services, but are degraded to make way for marketed products. Such is the case, for example, of wetlands. Wetlands are favourite sites for housing because of the clear view they offer. They are also sites for mangrove trees, particularly sturdy trees used in many countries as construction material or for fuel. Many wetlands are dried out to build houses, or mangrove trees cut. The functions performed by natural mangrove forests (e.g. clean waste water) are not included in the GDP, but when wetlands are transformed (e.g. into housing), this increases the GDP.

Damages caused by pollution (of air and water) have a negative impact on the GDP only if it negatively impacts productivity. More often than not, the negative effects of pollution are addressed through further expenditure, which increases GDP. Similarly, when ecosystem services are degraded, they may need to be restored (for example a forest may need to be replanted to reduce the risk of landslides), or replaced by man-made infrastructure (for example, a wall may need to be built to replace the coastline protection functions that were lost when a wetland was built on, or a sewage treatment plant may need to be built to replace the water purification functions of a wetland). Both restoration and replacement add value to the GDP, but do not add to welfare.

Since environmental services are not accounted for in the GDP, it is not known how much of the growth in GDP is the result of replacing degraded environmental services. Omitting the positive contribution that a healthy environment provides to the economy and to welfare, and presenting the cost of mending environmental degradation as an improvement of the economy (or current income) violate basic accounting principles (Cobb, Halstead and Rowe, 1995). Herman Daly presented this violation of basic accounting principles as 'the current national accounting system treats the earth as a business in liquidation' (cited in Cobb, Halstead and Rowe, 1995: 10).

5. The GDP does not encourage the preservation of income-generating natural capital

According to Hicks (1946), income is the maximum amount that can be consumed over a specific period without undermining the capacity to produce and consume the same amount in future periods. Lawn and Clarke (2008b) illustrate this idea through the example of a timber plantation. Given 1000 m³ of timber available during the first year, a regeneration rate of 5 per cent a year would amount to the regeneration of 50 m³ of timber (1000 m³ × 0.05). However, if 100 m³ of timber were to be extracted, exceeding the rate of timber regeneration, only 950 m³ of timber would be left at the end of the first year (1000 m³ + 50 m³ − 100 m³). The amount of timber regenerated the next year would equate to 47.5 m³ (950 m³ × 0.05). If extraction continued at the same rate, the resultant amount of timber at the end of the second year would be 897.5 m³ (950 m³ + 47.5 m³ − 100 m³). According to Hicks' definition of income, only 50 m³ of timber during the first year and 47.5 m³ during the second year could be considered income, while the additional 50 m³ and 52.5 m³ harvested, respectively, should be seen as capital depletion.

GDP violates this definition of income, as it does not subtract the costs of the depletion of natural capital that may accompany an increase in man-made capital (manufactured products). To maintain constant levels of man-made capital consumption, it is necessary to maintain the level of extraction of low-entropy raw material (such as timber), below or equal to their rate of regeneration. Translating this back to the GDP leads to the conclusion that a portion of GDP must always be set aside to replace depreciated and depleted capital, rather than used for current consumption (Lawn and Clarke, 2008b). A further portion of the GDP would also have to be set aside to cover the negative side effects of economic development on the environment (such as air pollution and resource depletion), which results in a reduction, or deterioration, of welfare. Natural capital depletion is not accounted for in the GDP (but is indeed shown in the GPI, see Chapter 4) because, as mentioned above, improved extraction and production technologies hide the effects of depletion.

Socioeconomic problems with the GDP

1. The GDP does not capture income inequalities

GDP is an aggregate, territory-wide measurement, which fails to consider the income inequalities in a country. Failure to take into consideration income inequalities in the calculation of GDP overstates the wealth of a large proportion of people. When a small percentage of the population owns a large proportion of the national income, GDP figures give a skewed picture of a country's standard of living. As Talberth, Cobb and Slattery (2007: 8) state, 'when growth is concentrated in the wealthiest income brackets it counts less towards improving overall economic welfare because the social benefits of increases in conspicuous consumption by the wealthy are less beneficial than increases in spending by those less well off'. The marginal benefits enjoyed from increases in consumption by a rich family are smaller than those enjoyed by an equal increase by a poorer family (Lawn, 2003; Lawn and Clarke, 2008b; Lawn and Clarke, 2010; Berik and Gaddis, 2011).

The Gini coefficient measures the equality of a nation's income distribution, with 1 (or 100 per cent) representing maximum inequality (where one person has all the income) and 0 representing maximum equality (where all citizens have the same). There are examples of countries in which the Gini coefficient rises (a sign of increasing inequality) when GDP per capita rises, and others in which the Gini coefficient drops (a sign of increasing equality) when GDP per capita rises. Bulgaria is an example of the former scenario, while Brazil is an example of the latter. Essentially, this suggests that there is no clear relationship between the GDP and changes in the Gini coefficient (even though in theory a rise in GDP should foster a more equal society). However, there is strong empirical evidence (Hayes *et al.*, 2014; Hsing, 2005), which indicates that income disparity leads to decreasing worker productivity and increasing social unrest, which calls for policies to decrease inequality.

Inequality in Hong Kong and Singapore is very large and growing. According to the Census and Statistics Department, the Gini coefficient of Hong Kong has constantly been increasing over the last decades from 0.43 in 1971 to 0.451 in

1981, 0.476 in 1991, 0.525 in 2001, and 0.537 in 2011. The situation of Singapore is slightly better, although Singapore's Gini coefficient also increased, from 0.454 in 2001 to 0.472 in 2011 (with a slight drop in 2013, to 0.463, or 0.412 after government transfers and taxes – the lowest since it was first calculated in 2000). A note of caution has to be made, since subsidized housing (specially in the Hong Kong case) helps the poorest strata, and therefore the consumption-levels Gini coefficient is not as high. However, it is clear that inequality has been increasing over the last decades (specially in Hong Kong), a pattern that GDP figures do not show, and that average GDP numbers are not very informative when the inequality is so large.

2. The GDP ignores non-marketed products

The GDP ignores the role that non-marketed products play in the lives of people, only covering market transactions, even though the former contribute to individual welfare. This is relevant to both developed and developing countries, but in general the poorer the country is, the more non-marketed (in particular self-produced) products people consume. For example, subsistence forest-dwelling farmers use very little cash, as they grow the rice they eat, gather in the forest the food to supplement their rice diet, and build their houses with timber from the forest. When people make a transition from the informal to the formal economy, the increase in GDP that results from that transition tends to severely overestimate the changing level of consumption, which partly consists in simply replacing non-marketed with marketed products, perhaps with little improvements in the standard of living. Thus, a subsistence farmer who shifts from producing food crop for himself to producing cash crops and buying food in the market will suddenly contribute to GDP, but may not actually see his standard of living improving.

3. The GDP ignores non-marketed labour services

The GDP ignores labour that is not paid, such as volunteer labour and household labour. This non-marketed labour contributes to the economy, and also improves general welfare, by bringing people together in cooperative initiatives at the societal and family levels. These services potentially yield additional indirect benefits, such as the effective use of free labour and the allocation for other purposes of money that might otherwise have been spent on wages (Lawn, 2003; Lawn and Clarke, 2008b; Lawn and Clarke, 2010; Berik and Gaddis, 2011). All in all, these unaccounted for non-market labour services likely make a great contribution to welfare. In economies that are expanding, household labour often joins the labour force, both because prices are increasing and additional income is required (especially for poor households), and because of higher expectations by household members in terms of consumption of marketed products. When household labour joins the workforce, domestic helpers may be hired to replace household labour, as is common in Hong Kong and Singapore. Both activities increase a country's GDP, but their effects on welfare remains unclear.

4. The GDP ignores the costs of social ills

The GDP fails to account for the social cost of unemployment and underemployment, overwork and loss of leisure time, family breakdown and divorce, criminality, and other social ills, since it only considers the associated economic costs (such as cost of policing, or the legal costs). The welfare implications of these costs remain hidden under a concept of welfare based exclusively on the growth of economic activities, such as the GDP. These costs are damaging to human productivity and well-being, and reflect the extent of a nation's social disunity or dysfunctionality (Lawn, 2003; Lawn and Clarke, 2008b; Lawn and Clarke, 2010; Berik and Gaddis, 2011). For example, unemployment and underemployment reflect the ineffective use of labour, which undermines potential income and can lead to social disharmony, which may result in increased crime rates and divorce rates.

Furthermore, social malaise and deterioration come at a high price, since the resources spent on combating these 'side effects' of inequality (e.g. punitive, legal and social services) are used to tackle preventable costs, instead of creating opportunities for raising the standard of living (Lawn, 2003; Lawn and Clarke, 2008b; Lawn and Clarke, 2010; Berik and Gaddis, 2011). From the perspective of the GDP, such societal costs are taken as additions to economic production and services, and, thus, they are mistakenly interpreted to add to welfare.

5. The GDP ignores external debt

Foreign debt is not included in GDP estimates. Yet, foreign debt can have serious implications on a nation's ownership of its welfare-yielding assets. These may include the ineffective allocation of resources, which are diverted towards repaying debt instead of maintaining man-made and natural capital, and producing goods (Lawn, 2003; Lawn and Clarke, 2008b; Lawn and Clarke, 2010; Berik and Gaddis, 2011). According to Lawn and Clarke (2008a), external debt has been recognized as a major hindrance to the sustainable development of countries because it paves the way for the unsustainable exploitation of natural resources. As mentioned above, one must keep in mind that when the rate of exploitation of natural resources exceeds the rate of regeneration, it undermines the resources necessary for production and consumption, thereby causing a decline in both man-made and natural capital (Lawn and Clarke, 2008b; Lawn and Clarke, 2010; Berik and Gaddis, 2011).

6. The GDP ignores defensive or rehabilitative expenditure

A portion of GDP must also be directed to defensive (e.g. against flooding, crime) or rehabilitative purposes to maintain people's productivity, as well as preventive and responsive healthcare services and various types of insurance policies (Lawn and Clarke, 2008b). The GDP counts such 'defensive' expenditures as benefits rather than as costs, even though they mostly do not contribute to well-being.

7. The timeframe of benefits from services and capital investments are ignored

In the calculation of GDP, the total sum of consumption expenditure on all types of goods and services is accounted for immediately during the current accounting period. This inherently assumes that all benefits from expenditure (be it for consumer durables or publicly provided infrastructure) are enjoyed only during the year of purchase and are subsequently lost (Lawn, 2003; Lawn and Clarke, 2008b; Lawn and Clarke, 2010; Berik and Gaddis, 2011). Unlike food, which is completely used up when eaten and therefore gives only 'instantaneous' service, durable goods, whether it is public investments (such as roads) or private investments (such as factories or machinery) provide services throughout the lifespan of the products. Measurements of well-being should also reflect the continuous psychic income (see Chapter 3) these products provide. However, GDP accounts for the benefits from expenditures only during the year such expenditures take place, and completely disregards the benefits during the following years. Thus, GDP is inflated in the year of expenditure and deflated during subsequent years (Lawn, 2003; Lawn and Clarke, 2008b; Lawn and Clarke, 2010; Berik and Gaddis, 2011).

Why is the GDP still used?

Most economists acknowledge the problems we just reviewed. Yet, the GDP is still being used as the predominant economic indicator. Indeed, most professional economists, teachers of economics, politicians (irrespective of political affiliation), policy-makers, and the media, focus on GDP figures, and call for continuous GDP growth, seemingly unaware of the criticisms being made of the GDP. Economists address this apparent paradox by acknowledging its weaknesses, but saying that these weaknesses do not mean that the GDP should be abandoned. The most common argument made in favour of continuing using the GDP is that there is a positive correlation between the GDP and other quality of life indicators such as infant mortality, life expectancy, adult literacy rate, civil and political liberties, and others (van den Bergh, 2009; Kunze, 2014; Lomborg, 2001). For example, in 1962 Arthur Okun, an economist for US President John F. Kennedy's Council of Economic Advisers, posited that for every three-point rise in GDP, unemployment would fall one percentage point (now known as Okun's Law). There are, nonetheless, a number of problems with these observations.

First, although it is true that there is a positive correlation between the GDP and indicators such as infant mortality, life expectancy, and adult literacy rate for a specific period of time, there is strong evidence that the indicators that positively correlate with the GDP do not necessarily improve beyond a given GDP level, or at least not at the same rate. There are also examples of health indicators negatively correlating with GDP growth. Tapia Granados (2012) studied the relation between GDP and health progress indicators (e.g. infant mortality rates, life expectancy) in England and Wales for a period of over 160 years. His study found a negative relation between GDP growth and health progress: 'the lower the rate of growth of

the economy, the greater the annual increase in LEB for both males and females.'
Moreover, he reported that the effect in such a negative relationship is stronger
between 1900 and 1950 (years marked by two world wars and the Great Depression)
than in 1950 to 2000 (years characterized by the growth of the welfare state in
Europe and the US), while being quite weak during the nineteenth century (Tapia
Granados, 2012: 688). He goes even further, stating that his 'results add to an
emerging consensus that mortality rates drop faster during recessions than during
expansions' (Tapia Granados, 2012: 689). Hence, beyond a certain income level,
the relationship between GDP growth and welfare weakens considerably.

Second, while it can be expected that indicators such as education and life
expectancy are correlated with GDP, since higher incomes allow for more
investment in education, and a better diet, there are other indicators which are
likely to be negatively correlated with the GDP. This is the case, for example, of
air quality, free time, work stress, and congestion, among others (Ordás, Valente
and Stengos, 2011; Aristotelous, 2014).

Third, there are often differences among income ranges, with some income
ranges experiencing a negative correlation, between particular welfare indicators
and aggregate, national, GDP growth. This may be the case, for example, when
economic growth causes prices to rise, but only a few individuals, within particular
income ranges, receive a compensating income increase. In this case, the main
limitation of the GDP is due to the fact that the GDP is an aggregate economic
indicator that ignores inequality, and that not all groups in a society will equally
benefit from economic growth, making average growth rates misleading.

Fourth, a positive correlation between alternative indicators and GDP is not
itself a proof of causation (Avendano, 2012; Hansen, 2012). Although an
argument can be made that better education leads to higher GDP, which leads to
better education in a positive feedback loop, there are also countries where such
positive correlations are not present. For example, there are countries in which
longer life expectancy and better education are not the product of similar increases
in GDP (Cuba is one example).

These arguments against the close correlation of aggregate indicators and the
GDP are supported by an extensive study of Easterly's (1999), which involved a
panel dataset of 81 indicators covering up to four time periods (1960, 1970, 1980,
and 1990) and seven subjects: 1) Individual rights and democracy; 2) Political
instability and war; 3) Education; 4) Health; 5) Transport and communications; 6)
Inequality across class and gender; and 7) 'Bads'. Using three different methods
of analysis, Easterly (1999) concluded that income per capita had an impact on
the quality of life that was 'significant, positive, and more important than
exogenous shifts' (p. 239) for only 32, 10, and 6 out of 81 (in the last case 69)
indicators. Easterly (1999) speculated that the result may be due to '(1) the long
and variable lags [...] between growth and changes in the quality of life, and (2)
the possibility that global socioeconomic progress is more important than home
country growth for many quality of life indicators' (p. 239). Regardless of the
reason for the lack of relationship, it is clear that there is no clear causal
relationship between the GDP and aggregate indicators of welfare.

The GDP has considerable impact on policies

Some proponents of the GDP indicator contend that the GDP has actually little influence in government policies. In reality, GDP figures play a central role in political discourse. Indeed, the failures of the GDP, mentioned in the previous section, would not be so important if GDP figures were not so influential in determining government policies. National and Supranational Economic agencies consider GDP information as vital information to explain, understand or predict the impact of economic policies. People celebrate when GDP growth figures are high or higher than expected, and show concern when GDP growth figures are low. When GDP figures register a decline, it becomes the focus of concern for the media and financial markets. Politicians and central banks respond with measures to raise the GDP. The importance of the GDP is promoted by the media, which informs about national GDP figures on a regular basis and provides cross-country comparisons. When the GDP is cited by news networks, reported by central banks, governments, international agencies and the business community on a regular basis, people take for granted that it provides important information, and do not question its usefulness. The influence of GDP information on the macroeconomic policies is so large that we sometimes forget how hard it is to accurately equate the sum of all goods and services produced in a country to general well-being.

The importance of education

Van den Bergh (2009) claimed that the widespread acceptance of GDP without much criticism from economists and students of economics could be attributed to conformism, docility, socialization and imitation. Docility helps in the learning process. Humans learn by absorbing ideas and information from others in an uncritical way (Simon, 1990). Being critical towards information obtained through social interactions, in particular from parents and teachers, would make the rapid learning and accumulation of knowledge more difficult. However, in turn that docility hampers people's ability to criticize or reject information obtained through social channels (Simon, 1990).

Colander and Klamer (1987) examined the responses of students in six top ranking graduate programmes in economics. The uniformity of responses revealed that the students were subject to a 'field of study socialization process'. Almost 20 years later, Colander (2005) investigated the same theme and found the same responses. When explaining this socialization process, Colander (2005) states that:

> Individuals are not born as economists; they are molded through formal and informal training. This training shapes the way they approach problems, process information and carry out research, which in turn influences the policies they favor and the role they play in society. [...] In many ways, the replicator dynamics of graduate school play a larger role in determining

economists' methodology and approach than all the myriad papers written about methodology.

(p. 175, in Van den Bergh, 2009)

Most of the time, economic education ignores the criticism of GDP as an indicator of welfare and reinforces the widespread belief in the importance of, and necessity to focus on, the GDP. For instance, most economic textbooks used by undergraduate students ignore critiques of GDP as an indicator of welfare when they discuss GDP growth. It is only in the final pages of the book that authors address the criticisms, but to a lesser extent and not without a degree of ambiguity. In many cases limitations are not even discussed: Gregory Mankiw's *Principles of Macroeconomics* (2009) dedicated over 30 pages to the construction and usage of the GDP, yet failed to address any single limitation about either its calculation or its role in economic theory. Similarly, David Weil's *Economic Growth* (2005) addressed the question: 'Will growth make us happy?' but failed to provide any clear answer and simply stated that, 'Income is not the only determinant of happiness, but clearly happiness rises with income...' and 'Thus, although growth will not make us as happy as we expect it to, it will still make us happier than we would be if there were no growth' (pp. 508, 510). As discussed in the previous pages and in Chapter 3, such accepted disjunctions are forced and valid only under certain conditions, which invalidate them as a general statement.

Should the GDP be abandoned?

As discussed in the previous paragraphs, the arguments made in favour of using GDP related figures to evaluate a country's development can be seriously misleading. Many politicians, economists and international agencies have failed to recognize this serious information failure associated with the GDP and have shown apathy towards getting it changed. In such a scenario, it would be helpful to know how a world without GDP-driven policies would look. Would it be better, or would it be worse? What would ignoring GDP information (and hence an emphasis on economic growth) mean for economic policies?

Economists contend that we should not abandon the GDP as long as good alternative indicators are not available. Yet, as the next chapter shows, many indicators are available that complement or supplement GDP figures. Even though none of them is perfect, we believe that they are more useful than the GDP in directing economic policy. In particular, we believe that the GPI addresses many of the concerns discussed, and – importantly – it reveals the point beyond which further economic growth is undesirable because the environmental and social costs that accompany economic growth outweigh the benefits of economic growth, and welfare decreases. If we give the social and environmental costs a negative sign, instead of a positive sign as GDP measurement does, we can conclude that in many rich countries, in the last decades, we have had negative economic growth (see Chapter 7).

Abandoning the GDP does not mean that economic growth would be forsaken, but that we would aim at improvements in welfare, using relevant and measurable indicators that tell us whether welfare has indeed improved, instead of aiming at economic growth alone, as the exclusive albeit indirect way to improve welfare. Abolishing the exclusive focus on GDP growth and our love for everlasting GDP growth rates, will allow for welfare metrics to gain in relevance and usage both within and outside academic circles. This would be a drastic change from the present.

If economists, politicians, policy-makers and consumers ignore the GDP, much of the economic behaviour that results from misleading GDP information might be averted. This will automatically reduce the resistance against policies aimed at increasing social welfare at the cost of GDP growth, paving the way for policies that aim at improving social welfare rather than GDP. Another benefit of ignoring GDP information is that it will create less panic responses and economic instability. The absence of GDP information may also give rise to new, more pragmatic approaches for developing countries which are transitioning into a formal economy. International institutions such as the World Bank and the UNDP have already taken steps in that direction, with indicators such as the Human Development Index (HDI, see Chapter 2) but have not yet totally discarded GDP. The Steady State Economy (see Chapter 8) offers an alternative set of policies that may be used to organize the economy of a country that does not attempt to maximize economic output as the ultimate goal of economic activities.

Conclusions

In this chapter we examined the theoretical and empirical problems of the GDP, and showed how in spite of all its imperfections, the GDP has remained the most commonly referred to indicator, mistakenly used also as a measure of economic development, social progress, and welfare. Its role in economic and political discourse is extremely important, and yet totally unwarranted. It seems that the vast majority of economists, journalists, civil servants and politicians do not ask themselves what the GDP shows, how it is calculated, what growths and shrinkages mean, and what the consequences of growth and shrinkage are. They simply look at the GDP as the most important indicator, and think that it needs to grow at all costs. The support for the GDP indicator turns out to be rather dogmatic, perhaps due to habitude, rather then the result of a reasoned choice.

We looked at the reasons for economists to continue using the GDP indicator. Many economists accept the criticisms of the GDP, but dismiss them, based on two arguments. First, they assert that either these criticisms do not actually weaken the usefulness of the GDP as an indicator. We have discussed the most common arguments put forward in favour of maintaining GDP, and shown their weakness. Second, apologetics of the GDP maintain that the importance of the measurement as a guide of economic policy is actually rather modest, the GDP being only one of the many indicators politicians look at when devising economic policies. Even though it is obviously difficult to prove that one of the main concerns of politicians is achieving high rates of GDP growth, its importance in public discourse tend to indicate that it is indeed considered a very important indicator by economists, politicians, news networks, and the population. We can conclude that the GDP 'represents a serious information failure: it suffers from many shortcomings and has a large influence on socioeconomic reality' (van den Bergh, 2009: 127). It would be best if a large number of economists pleaded together to abandon the GDP. Economists play a central role in maintaining the centrality of the GDP indicator, because of the voice they are given in news networks and policy formulation. However, as economists go through a process of indoctrination in their training, and the whole discipline rests on the importance of economic growth, it is unlikely that they would renounce the GDP any time soon. What is needed is to rethink the role of economic growth and GDP-related figures in the education given to economists. A first step should be to recognize the limitations and problems of the GDP.

Ignoring the GDP does not mean that the government should not aim at encouraging economic growth. It simply means that the government should not engage in policies with the *ultimate objective* of increasing economic output. The goal of government policies should be that of increasing social welfare rather than economic output. If economic output increases when welfare is improved, that is not an unhappy occurrence, but it should not be the objective of economic policy.

References

Anielski, M. (2001). *Measuring the Sustainability of Nations: The Genuine Progress Indicator system of Sustainable Well-being Accounts*. The Fourth Biennial Conference of the Canadian Society for Ecological Economics: Ecological Sustainability of the Global Market Place, Montreal.

Anielski, M. and Soskolne, C. (2002). Genuine progress indicator accounting: relating ecological integrity to human health and well-being. In: Miller, P. and Westra, L. (Eds), *Just ecological integrity: the ethics of maintaining planetary life*. Lanham, MD: Rowman and Littlefield Publisher Inc.

Aristotelous, K. (2014). Economic liberalization and environmental degradation. *International Journal of Ecology & Development™, 27*(1), 64–76.

Avendano, M. (2012). Correlation or causation? Income inequality and infant mortality in fixed effects models in the period 1960–2008 in 34 OECD countries. *Social Science & Medicine, 75*(4), 754–760.

Berik, G. and Gaddis, E. (2011). The Utah Genuine Progress Indicator, 1990 to 2007: A Report to the People of Utah. Available online: www.utahpop.org/gpi.html

Cobb, C., Halstead, T. and Rowe, J. (1995). If the GDP is up, why is America down?. *The Atlantic Monthly, 276*(4), 59–79.

Colander, D. (2005). The making of an economist redux. *Journal of Economic Perspectives, 19*(1), 175–198.

Colander, D. and Klamer, A. (1987). The making of an economist. *Journal of Economic Perspectives, 1*(2), 95–111.

Commission on Growth and Development. (2008). The Growth Report strategies for sustained growth and inclusive development. *The World Bank*. Retrieved from: www.ycsg.yale.edu/center/forms/growthReport.pdf

Costanza, R., Hart, M., Talberth, J. and Posner, S. (2009). *Beyond GDP: The need for new measures of progress*. Boston: Boston University.

Costanza, R., d'Arge, R., de Groot, R., Farber, S., Grasso, M., Hannor, B., Limburg, K., Naeem, S., O'Neill, R., Paruelo, J., Rasins, R., Sutton, P. and van den Belt, M. (1997). The Value of the world's ecosystem services and natural capital. *Nature, 387*, 253–260.

Davidson, M. D. (2013). On the relation between ecosystem services, intrinsic value, existence value and economic valuation. *Ecological Economics, 95*, 171–177.

Easterly, W. (1999). Life during growth. *Journal of Economic Growth*, 4, 239–276.

Geostat (2011). GDP calculation methodology. *Geostat.* Retrieved on 25 November 2011 from: www.geostat.ge/cms/site_images/_files/english/methodology/GDP%20Brief%20 Methodology%20ENG.pdf.

Hamilton, C. (1997). The Genuine Progress Indicator: A new index of changes in well-being in Australia. The Australian Institute, Australia.

Hansen, C. W. (2012). The relation between wealth and health: Evidence from a world panel of countries. *Economics Letters, 115*(2), 175–176.

Hayes, K. J., Slottje, D. J., Nieswiadomy, M., Redfearn, M. and Wolff, E. N. (2014). Productivity and income inequality growth rates in the United States. In: Bergstrand, J. H., Cosimano, T. F., Houck, J. W. and Sheehan, R. G. (Eds) *The changing distribution of income in an open U.S. economy.* Amsterdam: Elsevier.

Hicks, J. (1946). *Value and Capital* (2nd edn). London: Clarendon.

Hsing, Y. (2005). Economic growth and income inequality: the case of the US. *International Journal of Social Economics, 32*(7), 639–647.

Kunze, L. (2014). Life expectancy and economic growth. *Journal of Macroeconomics, 39*, 54–65.

Kuznets, S. (1934). Report to the United States Congress. Washington DC.

Lawn, P. (2003). A theoretical foundation to support the Index of Sustainable Economic Welfare, Genuine Progress Indicator, and other related indexes. *Ecological Economics, 44*, 105–118.

Lawn, P. and Clarke, M. (2008a). An introduction to the Asia-Pacific region. In: Lawn, P. and Clarke, M. (Eds), *Sustainable welfare in the Asia-Pacific: studies using the genuine progress indicator* (pp. 3–34). Cheltenham: Edward Elgar Publishing Ltd.

Lawn, P. and Clarke, M. (2008b). Why is Gross Domestic Product an inadequate indicator of sustainable welfare. In: Lawn, P and Clarke, M. (Eds), *Sustainable welfare in the Asia-Pacific: studies using the Genuine Progress Indicator* (pp. 35–46). Cheltenham: Edward Elgar Publishing Ltd.

Lawn, P. and Clarke, M. (2010). The end of economic growth? A contracting threshold hypothesis. *Ecological Economics, 69*, 2213–2223.

Lomborg, B. (2001). *The Skeptical Environmentalist: Measuring the real state of the world* (12th edn). Cambridge: Cambridge University Press.

Mankiw, G. N. (2009). *Principles of Macroeconomics* (7th edn), Ohio: South-Western Cengage Learning.

McCulla, S. H. and Smith, S. (2007). Measuring the Economy: A primer on GDP and the National Income and Product Accounts. Bureau of Economic Analysis, US Department of Commerce.

Ordás Criado, C., Valente, S. and Stengos, T. (2011). Growth and pollution convergence: Theory and evidence. *Journal of Environmental Economics and Management, 62*(2), 199–214.

Simon, H. A. (1990). A mechanism for social selection and successful altruism. *Science*, 250(4988), 1665–1668.

Stigliz, J. E., Sen, A. and Fitoussi, J. P. (2009). Report by the Commission on the Measurement of Economic Performance and Social Progress. Retrieved on 14 November 2014 from www.stiglitz-sen-fitoussi.fr/documents/rapport_anglais.pdf.

Stockhammer, E., Hochreiter, H., Obermayr, B. and Steiner, K. (1997). The index of sustainable economic welfare (ISEW) as an alternative to GDP in measuring economic welfare. The results of the Austrian (revised) ISEW calculation 1955–1992. *Ecological Economics, 21,* 19–34.

Talberth, J., Cobb, C. and Slattery, N. (2007). The Genuine Progress Indicator 2006. Retrieved on 14 November from www.lanecc.edu/sites/default/files/sustainability/talberth_cobb_slattery.pdf

Tapia Granados, J. A. (2012). Economic growth and health progress in England and Wales: 160 years of a changing relation. *Social Science & Medicine*, 74(5), 688–695.

Van den Bergh, J. C. J. M. (2009). The GDP Paradox. *Journal of Economic Psychology*, 30, 117–135.

Vu, Q. V. (2009). *GDP by production approach: A general introduction with emphasis on an integrated economic data collection framework.* Geneva: UNSD.

Weil, D. N. (2005). *Economic Growth.* Boston: Pearson/Addison-Wesley.

2 Alternative indicators to the Gross Domestic Product

The problems with the GDP discussed in the previous chapter led to the development of several alternative indicators that attempted to improve, supplement or replace the GDP. In this chapter we introduce some of such indicators, many of which were developed in an effort to measure sustainable well-being. We will also identify the deficiencies of these indicators. In the first section we introduce indicators which have economic data as the centrepiece, and can be said to adjust and improve the GDP. In the second section, we look at indicators which do not include economic data, or give it secondary importance, and therefore can be said to replace the GDP.

Indicators that adjust the GDP

Some alternative indicators of economic well-being use GDP as the foundation but make adjustments to it, in an attempt to address some of the limitations of the GDP discussed in the previous chapter. Some of these indicators are now reviewed.

Measure of Economic Welfare (MEW)

The Measure of Economic Welfare (MEW) is a measure that uses GDP as a foundation. It was first proposed in 1972 by William Nordhaus and James Tobin in their article *Is Growth Obsolete?* It was the first notable initiative that adjusted GDP and served as the first model for economic sustainability assessment. While developing the MEW, the authors in no way denied the importance of conventional national income accounts. However, they did question the usefulness of the Gross National Product (GNP, which at that time was more commonly used by policy-makers than the GDP) in evaluating economic welfare (Nordhaus and Tobin, 1972; Stewart, 1974). The authors argued that GDP was simply an index of production and not consumption, and hence actually violated the goal of economic activity (Nordhaus and Tobin, 1972). To construct their measure of welfare, Nordhaus and Tobin (1972) took national output as the foundation, but made several changes. First, they excluded 'regrettable necessities' (necessities that do not directly increase the economic welfare of households), such as expenditures on national security, prestige, or diplomacy. Second, they added the value of nonpaid household labour, illegal production, and leisure time. Third, they deducted environmental damage, or the costs of environmental pollution caused by industrial activity, known as the disamenity premium of urbanization.

The MEW made a significant contribution towards addressing the deficiencies of the GDP, as it outlined the limitations of the GDP, and its inability to measure economic welfare. However, despite including values of various economic activities not included in the national income accounts, the authors stated that MEW was primarily a broad measure of consumption and not economic welfare. Nordhaus and Tobin (1972) found it difficult to establish a correlation between welfare and consumption, as they realized the problems involved in trying to measure welfare (Stewart, 1974). The limitations of MEW, in terms of not being an effective index in computing welfare, can be best explained in the words of Nordhaus and Tobin: 'We are aiming for a consumption measure, but we cannot of course estimate how well individual and collective happiness is correlated with consumption' (Nordhaus and Tobin, 1972: 25). Even though the authors themselves term the MEW a primitive and experimental 'measure of economic welfare' (Nordhaus and Tobin, 1972: 4), it remains historically important as it served as a major source of inspiration for others who attempt to develop improved measures. For instance, the Income of Sustainable Economic Welfare (ISEW) developed by Herman Daly and John Cobb in the 1980s and the Genuine Progress Indicator (GPI) developed by Clifford Cobb were conceptually based on MEW (Goossens *et al.*, 2007).

Although the MEW contained aspects of sustainable development (such as the disamenity premium of urbanization), it was criticized for not making any adjustments to address environmental concerns affecting economic welfare, such as the depletion of natural resources (Hecht, 2002). Nordhaus and Tobin (1972) measured economic growth and welfare between 1929 and 1965 using the MEW. The authors found that while the growth rate of the GDP was consistently higher than that of the MEW, they followed a similar trend. This analysis led the authors to conclude that economic growth (as measured by the GDP) was closely related to economic welfare and therefore could be used as an indicator of well-being (Jackson and McBride, 2005). This led Nordhaus (1992 as cited in Jackson and McBride, 2005: 17) to examine the same question (*Is Growth Obsolete?*) from an environmental perspective. Nordhaus concluded that the differences between the MEW and the GDP were due to the decline in productivity growth and savings and not to the unsustainable use of natural resources. Nevertheless, despite making certain social, economic and environmental adjustments to the MEW, the latter cannot be considered an indicator of economic welfare, social well-being or quality of life (Jackson and McBride, 2005).

Green national accounting or Green GDP

Green national accounting or Green GDP, uses GDP as a foundation but adds the costs of environmental degradation (to air, water and soil) that result from economic activities, and the depletion of non-renewable resources. While there is a debate about who invented the Green GDP, the concept gained importance in 1992 at the United Nations Conference on Environment and Development (UNCED) held in Rio de Janeiro. There, Member States agreed on the importance to deal with the world's environmental problems, which required a national accounting system that better integrates the level of environmental degradation, and the conditions of the environment (Jinnan, Hongqiang and Fang, n.d.).

The Green GDP attempts to account for the loss of ecosystem services, degradation of the natural resources, pollution of soil, water and air, and in general damages to the environment by subtracting the costs to the environment from the overall GDP (Alfsen *et al.*, 2006). The Green GDP made significant contributions in correcting the GDP as it attempted to measure the growth of an economy while taking into account the harm that production does to the environment, which is not captured in the GDP (Goossens *et al.*, 2007). As a result, Green GDP has been at the forefront of efforts to integrate economic and environmental concerns and bring policy attention to the quality and sustainability of economic growth. Numerous attempts to calculate the Green GDP were made in various countries around the world (Costanza *et al.*, 2009). These attempts were largely based on subtracting the costs of natural resource depletion and pollution from GDP (Wu and Wu, 2010). However, none of these attempts resulted in regular reporting of Green GDP or succeeded in making Green GDP the hallmark indicator of economic fitness. This is because, since its inception, the concept of Green GDP has come in for a series of criticisms, mainly centred around the questions of precisely which cost items to deduct from the GDP and how to quantify the costs of these items when addressing environmental damage in monetary terms (Alfsen *et al.*, 2006; Goossens *et al.*, 2007). The inherent challenges in green GDP accounting are best described in The SEEA Handbook (UN, 2003). According to the handbook, 'there is no consensus on how 'green GDP' could be calculated and, in fact, still less consensus on whether it should be attempted at all' (p. 415). As a result, today, while Green GDP remains an index that raises awareness of sustainability concerns among policymakers and the general public, it is not widely used.

Genuine Savings (GS) (adjusted net savings)

In its twentieth World Development Report, the World Bank (1997) introduced 'Genuine Savings' (GS) to evaluate the national sustainability of countries around the world. Similarly to the two indicators introduced above, GS is also an indicator that uses GDP data as the starting point before adding and subtracting certain values (Everett and Wilks, 1999; Maro, 2007). GS aims to provide policymakers with a clear, relatively simple indicator, which they can use to evaluate the sustainability of their country's investment policies. Yet, the GS defines wealth more broadly than orthodox national accounts such as the GDP. To do so, it recalculates the national savings figures of a nation by deducting the cost of natural resources depletion (such as forests and water), pollution damages (including loss of welfare in the form of human sickness and health), and net borrowing. At the same time, the GS adds the current expenditure on education, treating it as saving rather than as consumption, as it increases a country's human capital (Everett and Wilks, 1999). In essence, the World Bank wanted to develop an index that differed from traditional national accounting systems, by including a range of assets that are important for development. Figure 2.1 shows the method that the World Bank uses for calculating the GS.

Genuine Savings figures aim to denote the rate at which national wealth, which includes human capital and natural capital, is being created or destroyed. The World Bank presents annual GS figures, as a percentage of the Gross National Income (GNI), in its World Development Indicators reports (Everett and Wilks, 1999). The main advantage of GS is that the World Bank presents the GS either as a positive or a negative figure where a negative figure means that a country is diverting from the path of sustainable development. A persistent negative GS implies that the total wealth of a nation is declining and the policies that the country has adopted are unsustainable (World Bank, 2005; Goossens *et al.*, 2007).

Although the Genuine Savings Indicator introduced by the World Bank makes it possible to answer a number of policy questions, important towards sustaining development, the indicator has been criticized on several grounds. First, the GS has been criticized for using GDP/economic growth as the core measure of development and progress (Maro, 2007; Goossens *et al.*, 2007). This is mainly because countries with high and growing GDP are more likely to obtain a strong and positive GS than countries with a low or shrinking GDP.

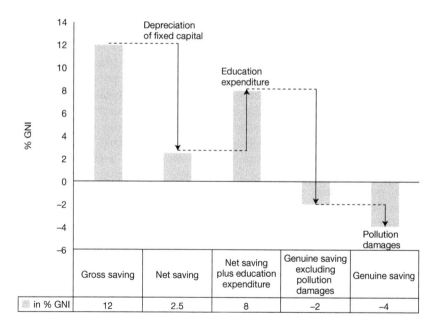

Figure 2.1 Calculation of Genuine Savings

Second, the GS was criticized for including investments in human capital (e.g. education) as net domestic savings rather than expenditure. In the World Bank's 2005 report 'Where is the Wealth of Nations? Measuring Capital for the 21st Century', the increased wealth in high-income countries was mainly due to the growth of intangible wealth, such as human capital. On the other hand, in poor countries the increased wealth was mostly due to the sale of natural resources, rather than investing in human capital, which is likely to result in lower welfare in the long run (Costanza *et al.*, 2009). Developed countries tend to invest more in education than non-developed countries, and so including investment in human capital in savings figures shows developed countries as more sustainable than less developed ones. This limitation of the GS failed to conceal inequality between regions and countries and made it an indicator to be used specially for country-level assessment of social, economic and environmental progress (Everett and Wilks, 1999; Goossens *et al.*, 2007).

Third, similarly to the Green GDP, Genuine Savings has also drawn flak for its inadequate methodologies to convert social and environmental variables into monetary values, and to estimate the cost of natural resource depletion and environmental pollution. Because of all these limitations, GS is used to measure sustainability and to determine the level of social, economic and environmental progress rather than to measure human welfare.

Sustainable Development Indicators (SDI) by the EU Sustainable Development Strategy (EU SDS)

Since the 1992 United Nations Conference on Environment and Development (UNCED), the European Union (EU) played a leading role in supporting the idea of sustainable development, and took various steps to integrate environmental concerns into its policies. The most notable step was the adoption of the EU SDS in 2001 as a response to the Gothenburg European Council meeting. The EU SDS required the commission to develop indicators at an appropriate level of detail to monitor progress with regards to the overall long-term quality of life and well-being of present and future generations (European Commission, 2011). The indicators were developed by Eurostat, with the help of a group of national experts known as the Task Force on Sustainable Development Indicators. The Commission first came up with a set of indicators in February 2005 and updated them in 2007. These indicators are structured as a three-level pyramid (Figure 2.2) where each level reflects the structure of the EU SDS (overall objectives, operational objectives, actions) and also responds to the user needs. The headline indicators have the highest communication value, and are complemented with contextual indicators. These contextual indicators are not directly used to monitor progress towards the achievement of the strategy's objectives but provide valuable background information for the analysis (European Commission, 2011; Endl and Sedlacko, 2012).

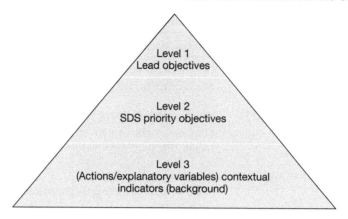

Figure 2.2 The SDI Pyramid

The Commission presented a set of 155 indicators in February 2005. Of more than one hundred indicators, ten were identified as headline indicators based on ten themes, reflecting the seven key challenges of the SDS, the important objective of economic prosperity, and guiding principles related to good governance (Table 2.1). These indicators also follow a gradient from the economic, through the social and environmental to the global and institutional dimensions (European Commission, 2011; Endl and Sedlacko, 2012).

Each of the headline indicators is divided into sub-themes to organize the set according to the operational objectives and actions of the EU SDS. The EU uses these SDIs to monitor the EU Sustainable Strategy and to evaluate its progress towards achieving sustainable development as per the targets and objectives defined in the EU SDS. These monitoring reports are published by the Eurostat every two years (European Commission, 2012).

The EU SDS discourse has been at the centre of controversy on how to best measure, monitor and assess progress towards sustainable development (Hametner and Steurer, 2007). A report released by the Environmental Audit Committee, questioned the inclusion of growth rate of real GDP per capita as a headline indicator. The Committee argued whether GDP should be an SDI and termed its relevance as doubtful especially in the view of the intention of the SDIs to capture 'inter-generational' rather than 'current', well-being (Environmental Audit Committee, 2012). By 2007, the European Commission uses its sustainable development indicators only for policy analysis and was still working towards launching a feasibility study on a well-being indicator.

Table 2.1 Breakdown of Sustainable Development Indicators (SDI)

Theme	Headline Indicator
Socioeconomic development	Growth rate of real GDP per capita
Sustainable consumption and production	Resource productivity
Social inclusion	People at-risk-of-poverty or social exclusion
Demographic changes	Employment rate of older workers
Public Health	Healthy life years and life expectancy at birth, by sex
Climate change and energy	Greenhouse gas emissions Share of renewable energy in gross final energy consumption Primary energy consumption
Sustainable transport	Energy consumption of transport relative to GDP
Natural resources	Common bird index Fish catches taken from stocks outside safe biological limits: Status of fish stocks managed by the EU in the North-East Atlantic
Global partnership	Official development assistance as share of gross national income
Good governance	No headline indicator

Source: European Commission, 2012

System of Economic Environmental Accounts (SEEA)

The SEEA was introduced jointly by the United Nations, the European Commission, the International Monetary Fund (IMF), the Organization for Economic Co-operation and Development (OECD) and the World Bank in the year 2003. Brouwer *et al.* (2013) define SEEA as 'a system for organizing statistical data for the derivation of coherent indicators and descriptive statistics to monitor the interactions between the economy and the state of the environment to better inform decision-making' (p. 70). The SEEA was mostly developed to measure the contribution that the environment makes to the economy and the impact that the economy has on the environment (Havinga, 2011). The SEEA comprises four categories of accounts: 1) flow accounts for pollution, energy and materials; 2) environmental protection and resource management expenditure accounts; 3) natural resource asset accounts; 4) valuation of non-market flow and environmentally adjusted aggregates (UN, 2003).

The main strength of the SEEA is that it is a single organized framework that attempts to integrate many of the different methods proposed for environmental accounting. For instance, the SEEA integrates monetary indicators (such as environmentally adjusted net domestic product, capital formation or value added) that measure sustainable economic activity and growth with physical indicators (such as flows and stocks, notably natural resource inputs and 'outputs' of pollutants and wastes). When linked to economic performance indicators, especially the GDP, these monetary indicators measure the environmental impact of economic growth, as a ratio of material intensity or resource productivity (Pinter, Hardi and Bartelmus, 2005).

The SEEA can be used to support policymaking. For instance, the SEEA can help understand the consequences of different policy options on the atmospheric, soil, and water pollution. It can also provide valuable 'information regarding environmentally related transactions such as taxes and subsidies to examine cost-recovery or polluter pays principles' (London Group on Environmental Accounting, 2007: 292). Despite the comprehensiveness of SEEA in recording the interaction between economic processes and the environment, it is merely a proposed methodology, and falls short of being an international statistical standard. Nevertheless, the United Nations Statistical Commission (UNSC) established the United Nations Committee of Experts on Environmental-Economic Accounting (UNCEEA) to elevate the SEEA to an international statistical standard, and advance its implementation (London Group on Environmental Accounting, 2007).

National Accounting Matrix including Environmental Accounts (NAMEA)

In the late 1990s, Statistics Netherlands developed the NAMEA. The NAMEA is a framework for documenting economic and environmental flows in a consistent way following the UN System of National Accounts (SNA) established in 1953. According to the European Environment Agency (EEA, 2007), which undertakes projects examining environmental pressures in selected European countries, the main components of NAMEA are 'conventional economic input-output matrices – national inventories of monetary flows between economic sectors and between them and final consumers. These inventories are then extended by adding information on material resource inputs to each sector and the pollutants they release back into the environment' (EEA, 2007: 1). This feature of the NAMEA serves as a useful tool to support policy design and analysis in the arena of sustainable consumption and production (SCP). It helps policymakers in approaching environmental issues both from the production and consumption side. It also allows identifying environmental hotspots in the system (Goossens *et al.*, 2007).

Although the NAMEA has been hailed for providing a comprehensive analysis of the production and consumption systems and focusing on environment-economic aspects, it is still criticized for its large data requirements, for its difficult methodology and for excluding social aspects. To date, the NAMEA approach has been adopted by Eurostat and implemented in many European countries. It is integrated into the SEEA and is used to support European and national policymaking in the area of SCP. The limited availability of data, however, remains a serious constraint of the NAMEA and needs to be addressed (Goossens *et al.*, 2007).

Indicators that replace the GDP

A number of indicators have been developed to address the need for indices that look at more than purely economic data. These indicators can be categorized as: 1) Composite indices, which include several indicators, including GDP, into a single measure. These include the Human Development Index (HDI) and the Gender-Related Development Index (GDI). 2) Indices that include social and environmental variables, instead of the GDP. These include the Physical Quality Life Index (PQLI), the Human Poverty Index (HPI), the Ecological Footprint (EF), the Happy Planet Index (HPI), the Environmental Sustainability Index (ESI), the Environment Performance Index (EPI), and the Millennium Development Goals (MDGs). We also introduce three local indices: the Quality of Life Index, developed by the Chinese University of Hong Kong, the Social Development Index (SDI) by the Hong Kong Council of Social Service, and the City Biodiversity Index (CBI) developed by the Singapore government.

Physical Quality Life Index (PQLI)

The first major attempt at generating a composite indicator of development occurred in the late 1970s when Morris and McAlpin developed the PQLI. In the PQLI, three indicators – life expectancy at age 1, infant mortality and literacy were used to form a simple composite index. The PQLI was calculated by obtaining the average of these three indicators. For each of the three indicators, countries were ranked and given a score, with the 'best' country being given a score of 100, and the 'worst' performing country being given a score of 0 (Morris and McAlpin, 1979).

Human Development Index (HDI)

During the 1990s, the PQLI was replaced by the HDI. In the Human Development Report (UNDP, 2006), the HDI is defined as 'a composite measure of three dimensions of human development: living a long and healthy life (measured by life expectancy), being educated (measured by adult literacy and enrolment at the primary, secondary and tertiary level) and having a decent standard of living (measured by purchasing power parity, PPP, income)' (p. 263). It was first proposed in 1990 by Amartya Sen and Mahbud ul Had (Sen, 2004). The idea of HDI is mostly based on the Physical Quality of Life Index (PQLI). Just like the PQLI, the HDI also uses three measures to generate an index, and two of the three measures are related – literacy and life expectancy. However, unlike PQLI, the HDI uses GDP per capita on a Purchasing Power Parity (PPP) basis rather than infant mortality to balance social measures of development with an economic measure. Similar to the PQLI, the calculation of the HDI involves ranking countries on a scale from 100 to 0, and taking an average of the three rankings. The HDI is expressed on a scale from 0 to 1 where countries with HDIs of 0.800 and above are classified as countries with high human development whereas countries with HDIs of 0.500 to 0.799 are considered as medium human development countries and countries with HDIs of below 0.500 are considered as countries with low human development (Codrington, 2005). The United Nations Development Program (UNDP) publishes the HDI of 177 countries in its annual Human Development Report. According to the UNDP, the main aim of the Human Development Report is to stimulate global, regional and national policy discussions on issues that are relevant to human development.

Since its inception, the HDI has been criticized on several grounds (e.g. Sagar and Najam, 1998). For instance, it has drawn flak for relying on traditional GDP and for ignoring ecological aspects of sustainability such as environmental degradation. The HDI has also been targeted for depending on data that are difficult to obtain, especially in less developed countries. Today, the HDI is the most commonly used composite indicators of human development but its methodological weaknesses have largely reduced its policy relevance. This lack of completeness has marked its failure as a universal measure of sustainable development.

Gender-Related Development Index (GDI)

The HDI succeeded in raising awareness of the concept of 'human development'. However, it was severely criticized for neglecting other aspects of human development such as inequalities between genders. As a result, in 1995 the United Nations, introduced a gender-related indicator in its Human Development Report (UNDP, 1995). This indicator was the GDI. The main objective of the GDI was to add a gender-sensitive dimension to the HDI. The GDI is defined as 'distribution-sensitive measure that accounts for the human development impact of existing gender gaps in the three components of the HDI, namely life expectancy, literacy and GDP per capita' (Klasen, 2006: 243). The GDI uses an 'inequality aversion' penalty, which creates a development score penalty for gender gaps in any of the three categories of the HDI. In terms of life expectancy, the GDI assumes that women will live an average of five years longer than men and considers income gaps in terms of actual earned income. The GDI on its own is not considered an independent measure of the gap between genders. It is only used in combination with the scores from the Human Development Index to assess the extent of gender inequality (Klasen and Schüler, 2011).

Since its inception in 1995, there has been a lot of debate surrounding the usefulness of GDI. For instance, it is often mistakenly interpreted as a measure of gender inequality when, in fact, it did not intend to be interpreted that way. Additionally, just like the HDI, it has been criticized for its conceptual problems, such as relying on data that is not always readily available in developing countries. These factors have made the GDI an index that is hard to calculate uniformly and internationally (Klasen and Schuler, 2011).

Human Poverty Index (HPI)

In 1997, the UNDP introduced the HPI to assess the standard of living in a country. The HPI concentrates on the deprivation of the three essential elements of human life reflected in the HDI, but to better reflect socioeconomic differences, as well as the widely different measures of deprivation, the HPI is derived separately for developing countries (HPI-1) and a group of selected high-income industrialized countries (HPI-2) (Anand and Sen, 1997). For developing countries, it includes information on: 1) life expectancy; 2) adult literacy rate; 3) access to water and underweight children. For developed countries, it includes data on: 1) life expectancy; 2) adult literacy skills; 3) income inequality; 4) long-term unemployment rate. The HPI aims to capture 'human poverty', i.e. failures to achieve the basic capabilities needed for human functioning rather than the level of consumption or income.

So far, the HPI has shown numerous strengths. For instance, the HPI indicators have directed greater policy attention on the standards of public services related to education and health. It has also captured some essential elements of life that relate to social and economic policies that go beyond income and consumption alone. But like the HDI, the HPI is also an incomplete measure. First, the data needed to construct the HPI suffers from various inconsistencies and gaps. Second, the HPI does not capture several other essential dimensions of human poverty, such as those related to participation, political freedom and cultural choices, which are some of the many essential dimensions of social well-being. Due to all these limitations, the HPI is not accepted as a universal measure of poverty or welfare, let alone of social well-being or quality of life (Parr, 2006; Krishnaji, 1997). As a result, the HPI is predominantly an indicator that can be used to assess areas that require policy attention, as opposed to an indicator that reflects quality of life and welfare (Parr, 2006).

Ecological Footprint (EF)

The EF was first proposed by Marthis Wackernagel and William Rees. Wackernagel and Rees (1996) define EF as 'an accounting tool that enables you to estimate the resource consumption and waste assimilation requirements of a defined human population or economy in terms of corresponding productive land area' (p. 9). The EF tracks humanity's demands on the biosphere by comparing humanity's consumption against the earth's regenerative capacity, or biocapacity, by calculating the land area that would be needed to produce the resources that an individual or population consumes, the area needed to assimilate waste (including the area of forest and oceans required to sequester CO_2 emitted from fossil fuels), and the area occupied by infrastructure (Wackernagle *et al.*, 2002; Goossens *et al.*, 2007).

To calculate the per capita ecological footprint or the number of global hectares demanded per person, all (both marine and inland) areas that support photosynthetic activity and biomass accumulation that can be used by humans are added up and then divided by the population. Marginal and non-productive areas such as deserts and arid regions, and other low productive surfaces are not included. The result obtained is presented in hectares per capita (Kitzes *et al.*, 2009; Lenzen and Murray, 2003).

Humanity's EF has exceeded earth's total biocapacity – an 'ecological overshoot' – since 1970 (Figure 2.3), and the deficit has kept increasing since then. Part of the problem is that the biocapacity (the ability of the earth to sustain consumption and recycle waste) has kept decreasing since the 1970s, the result of environmental degradation (including widespread deforestation). According to the latest figures, the earth's total biocapacity was 12 billion global hectares (gha) or 1.8 gha per person in 2008 whereas human's Ecological Footprint was 18.2 billion gha or 2.7 gha per person in the same year. This difference or overshoot indicates that it would take 1.5 years for the earth's ecological capacity to regenerate the demand that people put on the biosphere in one year, along with the land and water bodies needed for absorption of wastes (WWF, 2012). Resources can be harvested faster than they can be regenerated because in the past they were left to accumulate. This is a similar situation to a person choosing to consume more than he can finance through income, and using savings to support his lifestyle for four months every year.

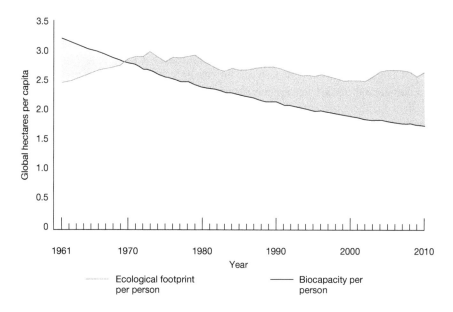

Figure 2.3 Trend in Ecological Footprint and biocapacity per capita between 1961 and 2010

Source: Global Footprint Network (2014)

Over the years, the EF has emerged as a standalone index of environmental sustainability and is widely used by scientists, businesses, government agencies and non-government organizations (Global Footprint Network, 2011). However, since its formulation, the EF has been criticized for not capturing every aspect of sustainability. For instance, the EF makes no assumption about technological progress, simply reflecting their actual influence on the current demand on the earth (Global Footprint Network, 2011). According to Kitzes *et al.* (2009), it is illogical to compare consumption level and biocapacity without taking into account technological progress, which is inevitable with time. This limitation of the EF can be best described in the words of Fiala (2008), when he states that, 'While the ecological footprint offers a simple and intuitive estimate of the production inputs for a given consumption level, it fails to address the sustainability of consumption that it was originally conceived to do' (p. 524).

Happy Planet Index (HPI)

The HPI is an index of environmental impact and human well-being. It was first proposed in 2006 by the New Economics Foundation (NEF), a charity founded in 1986 by the leaders of The Other Economic Summit (TOES) (Goossens *et al.*, 2007; NEF, 2012a). The HPI is a new measure of progress that focuses on sustainable well-being for all by measuring a country's ecological efficiency in delivering human well-being. The HPI is a composite of two objective indicators, life expectancy and ecological footprint per capita and one subjective indicator 'experienced well-being'. The data for 'life expectancy at birth' is obtained from the UN Human Development Reports for life expectancy, whereas the data for the Ecological footprint per capita is obtained from the Global Footprint Network. The data for the experienced well-being indicator is obtained by asking people how they themselves feel about their lives using a question called the 'Ladder of Life' from the Gallup World Poll. To measure the experienced well-being, the respondents are asked to imagine the step they would stand on a ladder, where 0 represents the worst possible life and 10 the best possible life (NEF, 2012a).

Measured on a scale from 0 to 100, the NEF set the global target for HPI at 89, based on attainable levels of well-being, life expectancy, and a reasonably-sized Ecological Footprint. The third and the latest Happy Planet Index report released by the NEF (2012b) that reported data for 151 countries showed that Costa Rica scored the highest HPI with 64.0; while the lowest HPI was 22.3 for Botswana (NEF, 2012b).

The HPI is considered an innovative index as it combines well-being and environmental aspects. Since it considers the end outcome of economic activity in the form of ecological sustainability, experienced well-being, and longevity, it goes beyond GDP and highlights the variety of factors that increase or decrease well-being. Furthermore, the method of calculating the index is simple and easily comprehensible by both the general public and the policymakers (Goossens *et al.*, 2007). However, although the HPI provides a useful insight of whether a society is heading in the right direction, some data gaps still remain, especially related to 'experienced well-being' and the environmental footprint. For instance, the HPI does not measure the infringement of human rights that negatively impact human well-being in some countries. Additionally, the HPI fails to establish a direct relationship between policy actions and experienced well-being. As Goossens *et al.* (2007) argue, experienced well-being or happiness is a very personal and subjective issue. This restricts the usefulness of the HPI as a measure for assessing the impact of policies. Finally, as the HPI includes Ecological Footprint, it is plagued with the latter's inherent limitations. Owing to these limitations, the NEF itself argues that the key value of the HPI lies in the fact that it captures an overall sense of how well a nation is doing, but that the HPI should be supplemented by other indicators (NEF, 2012b).

Environmental Sustainability Index (ESI)

The ESI was first launched in 1999 by Professor Daniel C. Esty along with the World Economic Forum's Global Leaders for Tomorrow Environment Task Force and Columbia University's Center for International Earth Science and Information Network (CIESIN). The ESI is a composite index measuring overall progress towards environmental sustainability. It tracks a diverse set of socioeconomic and environmental sustainability indicators at the national scale and is based on five components: environmental systems, reducing environmental stresses, reducing human vulnerability, social and institutional capacity and global stewardship, for a total of 21 indicators (Table 2.2) (Esty *et al.*, 2005).

The validity, interpretability and explanatory power of the ESI depends heavily on the quality and completeness of the input data. Unfortunately, the ESI suffers from various data uncertainties and methodological issues. In particular, the selection of the indicators and the equal weighting of the 21 indicators weaken the significance of the ESI. The limitations of the ESI has been recognized by Esty *et al.* (2006), who state that the ESI has been 'criticized for being overly broad – and not focused enough on current results to be useful as a policy guide. The concept of sustainability itself is partly at fault. Its comprehensive and long-term focus requires that attention be paid to natural resource endowments, past environmental performance, and the ability to change future pollution and resource use trajectories – as well as present environmental results' (p. 7). It is because of these criticisms of the ESI that the authors developed another index known as the Environmental Performance Index (EPI) to focus on countries' existing environmental performance within the context of sustainability.

Table 2.2 Components of the Environmental Sustainability Index (ESI)

Environmental Systems

1. Air Quality	4. Water Quality
2. Biodiversity	5. Water Quantity
3. Land	

Reducing Environmental Stress

6. Reducing Air Pollution	9. Reducing Waste & Consumption
7. Reducing Ecosystem Stress	Pressures
8. Reducing Population Pressure	10. Reducing Water Stress
	11. Natural Resource Management

Reducing Human Vulnerability

12. Environmental Health	14. Reducing Environment Related Natural
13. Basic Human Sustenance	Disaster Vulnerability

Social and Institutional Capacity

15. Environmental Governance	17. Private Sector Responsiveness
16. Eco-Efficiency	18. Science and Technology

Global Stewardship

19. Participation in International	21. Reducing Transboundary Environmental
Collaborative Efforts	Pressures
20. Greenhouse Gas Emissions	

Source: Esty *et al.* (2005)

Environmental Performance Index (EPI)

The EPI was developed to address the criticisms of the ESI, of not focusing sufficiently on present or current environmental conditions. The EPI focuses on countries' current environmental performance within the context of sustainability, and provides benchmarks for natural resource management results and current national pollution controls. It identifies specific targets for environmental performance and measures countries' achievements with respects to these goals (Esty *et al.*, 2006; Goossens *et al.*, 2007). To do so, the EPI centres on two comprehensive environmental objectives: 1) reducing environmental stresses on human health; and 2) promoting ecosystem vitality and sound resource management. The EPI gauges these objectives using 16 indicators in six policy categories: Environmental Health, Air Quality, Water Resources, Biodiversity and Habitat, Productive Natural Resources, and Sustainable Energy (Table 2.3) (Esty *et al.*, 2006).

Table 2.3 Components of the EPI

Objective	Policy Category	Indicator
Environmental Health		Urban Particulates
		Indoor Air Pollution
		Drinking Water
		Adequate Sanitation
		Child Mortality
Ecosystem Vitality and Natural Resource Management	Air Quality	Urban Particulates
		Regional Ozone
	Water Resources	Nitrogen Loading
		Water Consumption
	Biodiversity and Habitat	Wilderness Protection
		Ecoregion Protection
		Timber Harvest Rate
		Water Consumption
	Productive Natural Resources	Timber Harvest Rate
		Overfishing
		Agricultural Subsidies
	Sustainable Energy	Energy Efficiency
		Renewable Energy
		CO_2 per GDP

Source: Esty *et al.* (2006)

The EPI is calculated by converting the 16 indicators to a proximity-to-target measure with a theoretical range of zero to 100. The results are then presented in the form of cross-country comparisons. As a result, the EPI clarifies which governments are performing well in their approach towards environmental protection (Esty *et al.*, 2006). This EPI has become a valuable tool for national policymakers as it enables them to identify policy options, particularly in the context of environmental protection. However, in spite of these benefits to the national policymakers, the usefulness of the EPI is limited by data problems and methodological questions. To achieve its full potential, there is a need for better environmental data, which unfortunately remains a challenge, especially in developing countries.

Millennium Development Goals (MDGs)

In the Millennium Summit of the United Nations in the year 2000, 189 world leaders adopted the United Nations Millennium Declaration and officially established the MDGs to free people from extreme poverty and multiple deprivations by 2015 (UN, 2012). The goals are represented in Figure 2.4.

Forty-eight indicators were identified to measure progress achieved towards implementing these eight goals. These 48 indicators are aggregated at global, national and regional levels and used to evaluate progress over the period of 15 years from 1990 to 2015. Every year, the Secretary-General presents a report to the UN General Assembly on the progress achieved towards implementing the Declaration, based on the data of 48 selected indicators (Goossens *et al.*, 2007).

The main strength of the MDGs is that they form a blueprint agreed to by a large number of countries. Also, the goals are easily understood and the indicators can be easily linked to policy targets, while they can help formulate development assistance (Goossens *et al.*, 2007). However, the MDGs have been criticized for covering issues that are of concern only to a set of developing countries and not the entire global community. Additionally, the goals follow a top-down approach and exclude local authorities and stakeholders. Pinter, Hardi and Bartelmus (2005) argue that the MDGs are weak and imprecise especially on environmental sustainability as they ignore the sustainability of global ecological support systems. Thus, the MDGs are not universally applicable and acceptable, as they focus mostly on the priorities of the least developed countries and have a unique weakness in the environmental domain.

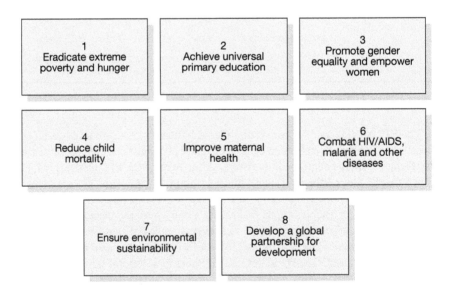

Figure 2.4 The Eight Millennium Development Goals For 2015

Source: UN (2012)

Hong Kong Quality of Life Index (HKQLI)

The HKQLI is the first of two indicators developed – and used – in Hong Kong. The objective of the index is to integrate the mutual influences between subjective feelings of individuals and objective environmental conditions (Chan, Kwan and Shek, 2005). The index consists of 21 indicators, which are grouped into three sub-indices, namely social, economic and environmental (Table. 2.4).

According to Chan, Kwan and Shek (2005) the main aims of the HKQLI are to measure and monitor the quality of life in Hong Kong, to enhance the quality of life by drawing the attention of the public to the issue, and to provide policymakers with useful statistics. Starting from the year 2003, using 2002 as the base year, the Index is released annually. The index for the year 2002 was set at 100. An increase means that the quality of life has improved, and a drop that it has worsened (CUHK, 2012a). Figure 2.5 shows that there has been a gradual improvement in the quality of life in Hong Kong from 2002 to 2006, and a drop afterwards.

Table 2.4 CUHK Hong Kong Quality of Life Index

Social sub-index

1. Standardized mortality rate (per 1,000 standard population)	6. General life satisfaction index
2. Life expectancy at birth (in year)	7. Press freedom index
3. Public expenditure on health as a proportion (in per cent) of the GDP	8. Press criticism index
4. Notification rate of notifiable infectious diseases (per 1,000 population)	9. Government performance index
5. Stress index	10. Overall crime rate (per 1,000 population)

Economic sub-index

11. Housing affordability ratio	15. Real wage index
12. Real rental index	16. Public expenditure on education as a proportion (in per cent) of the GDP
13. Unemployment rate	17. Age participation rate for first-degree programs and postgraduate programs in local universities (in per cent)
14. Index of current economic conditions	

Environmental sub-index

18. Air index	20. Noise index (per 1,000 population)
19. Water index	21. Recycle rate of municipal solid waste

Source: CUHK, 2012a

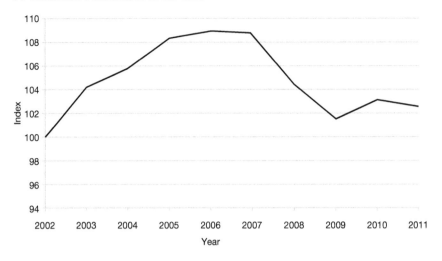

Figure 2.5 Quality of Life Index for Hong Kong

Source: CUHK (2012b)

The HKQLI is more comprehensive than other indices, since it incorporates both objective indicators based on measurements of observable conditions of social life, and subjective indices that measure individuals' subjective feelings. However, as Chan, Kwan and Shek (2005) argue, although the HKQLI provides useful references to policymakers, facilitates assessments of performances in various life domains, and identifies areas which require substantial improvement, it cannot be considered as an index that can measure individual well-being in Hong Kong. There are also serious methodological problems, with all components of the index given the same weight.

Social Development Index (SDI)

The SDI is a composite indicator developed by the Hong Kong Council of Social Services (HKCSS). According to Chan, Kwan and Shek (2005), the SDI is a composite indicator that 'relies on objective indicators, which are based on measurements of observable conditions of social life' (p. 264). The basic SDI includes 14 sectors of development (Table 2.5) involving 47 social, political and economic sub-indicators (HKCSS, 2002). Three types of indices are reported in the SDI: 1) the Weighted Social Development Index (WSDI), which describes the state and progress of social development at the societal level; 2) 14 Sub-indices, which describe the state and progress of development in particular domains (e.g. education, housing, environment, economy); 3) five Sub-indices, which describe the state and progress of development of particular social groups (women, children (0–14 years), youth, elderly, low income persons). The WSDI compares the overall social progress (bringing together the 14 sectors of development) of society as a whole, whereas the sub-indices give additional details for particular domains or social groups (HKCSS, 2002).

The first SDI report was released in the year 2002, with data from the year 2000 (SDI-2000). Since then, the SDI report has been updated bi-annually. Figure 2.6 shows a steady increase in social development in Hong Kong. However, sub-indices reveal a negative development of children, low-income families and youth (Land, Michalos and Sirgy, 2012). The SDI has contributed to the debate on the quality of life in Hong Kong, but has been criticized for not including the subjective feelings of individuals, which are important in assessing quality of life (Chan, Kwan and Shek, 2005).

Table 2.5 Social indicators used to form the SDI

14 Core sectors of development

1. Family Solidarity	8. Sports & Recreation
2. Health	9. Political Participation
3. Personal Safety	10. Strength of Civil Society
4. Economic	11. Housing
5. Environmental Quality	12. Education
6. Crime & Public Safety	13. Internationalization
7. Arts & Entertainment	14. Science & Technology

Source: HKCSS (2002)

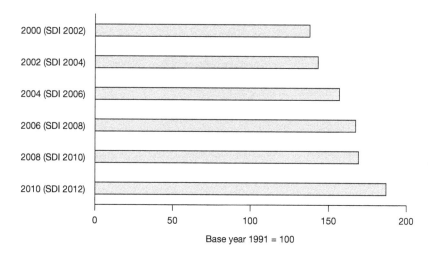

Figure 2.6 Standardized Weighted Social Development Index Scores

Source: HKCSS (2012)

City Biodiversity Index (CBI)

The CBI was first proposed in 2008 by Mah Bow Tan, then Minister for National Development of Singapore at the World Cities Summit of 2008 (CBD, 2010a). The aims of the CBI are to: 1) serve as a self-assessment tool and assist national governments in benchmarking biodiversity conservation efforts at the city level; 2) monitor progress in reducing the rate of biodiversity loss in urban ecosystems; 3) assist in measuring the ecological footprint of cities; and 4) make cities aware of important information gaps that exist concerning their biodiversity (CBD, 2009).

The index comprises 23 indicators (Table 2.6) grouped in three components: 1) native biodiversity, which focuses on the conservation of biodiversity; 2) ecosystem services provided by biodiversity, which includes carbon storage, water regulation and recreational and education services; 3) governance and management of native biodiversity within the city, which deals with budget allocation, public awareness and administrative programmes and other projects related to biodiversity (CBD, 2010b).

Table 2.6 Composition of the City Biodiversity Index (CBI)

THE CITY BIODIVERSITY INDEX		
Native biodiversity in the city (11 indicators)	*Ecosystem services provided by the native biodiversity in the city (5 indicators)*	*Governance and management of biodiversity in the city (7 indicators)*
1. Percentage of natural/ semi-natural areas	12. Freshwater services (Cost for cleaning water)	17. The budget allocated to biodiversity projects
2. The diversity of ecosystems	13. Carbon storage (No. of trees in the city)	18. No. of biodiversity projects and programs organized by the city annually
3. Fragmentation measures	14. Recreation and education services Nno. of visits/person/year)	19. Rules, regulations and policy
4. Native biodiversity in built-up areas		20. No. of institutions covering essential biodiversity-related functions
5–9. Native Species – Plants, Birds, Butterflies and 2 other species	15. Area of parks and protected areas/ population of the city	21. No. of inter-agencies coordinating
10. Percentage of protected areas	16. No. of educational visits to parks or nature reserves per year (under 16 years/year)	22. Existence of a consultation process (Incorporation of biodiversity into the school curriculum)
11. Proportion of invasive alien species		23. Existence of partnerships (No. of outreach programs and public awareness events

Source: Chan (2012)

Each of the 23 indicators is allotted a maximum score of four, with the total possible score of the index being 92 points. The year in which a city first starts the scoring is considered the base year (CBD, 2010b). The quantitative nature of the CBI allows cities to measure changes in the status of biodiversity over time. However, differences among ecological zones makes comparisons among cities difficult. Furthermore, some of the indicators are less meaningful in particular ecological zones (Rodricks, 2010). As of 2012, more than 50 cities are in various stages of applying the CBI, whereas more than 30 cities provided data on the indicators (Chan, 2012).

Conclusions

Doubts about the merits of the GDP as an indicator of welfare has brought about efforts to rethink how societal well-being, sustainability, and development should be measured. Indicators that adjust the GDP leave the GDP intact but complete it with a variety of indicators that show a comprehensive story of how society is doing, from a broader perspective than purely economic output. In contrast, dissatisfaction with giving any kind of relevance to economic output in assessing individual well-being has prompted the development of other indicators that completely ignore the GDP. These indicators measure well-being more directly, by aggregating a wide range of social and environmental items into a single index. For example, the Ecological Footprint completely ignores economic data to assess the sustainability of people's lifestyles.

Many indices discussed in this chapter have the problem of lacking a common unit to weigh the economic, social, and environmental factors they include. A commonly chosen solution is to give all components the same weight. For example, the HKQLI gives each of its 21 indices a weight of 4.76 in the base year, and then adjusts it yearly. This means that a drop of 10 per cent of one component (e.g. 'Life expectancy at birth') is compensated by an improvement of 10 per cent of another (e.g. 'Crime rate' or 'Recycle rate of municipal solid waste'), even though these indices deal with completely unrelated issues (CUHK, 2013). Clearly, this is methodologically unsatisfactory.

Some indices (such as the HKQLI) approach this problem by giving each component the same weight, and then adjusting it yearly a value of 100 to the base year, and then adjusting the value yearly. However, this means that a drop of 10 per cent of one component (e.g. 'Press criticism') is compensated by an improvement of 10 per cent of another (e.g. 'Housing affordability'), even though the two are unrelated. Another problem of many indices is that all components are given the same weight. This is particularly problematic when comparing economic, social and environmental items.

In the following chapter we introduce the GPI, which we believe addresses these two problems. The GPI gives an economic value to the economic, social and environmental components. This does not mean that everything is – or should be – reduced to a commercial entity, or replaceable by other monetized items. The economic value is simply a common unit of analysis that can simultaneously be used to weigh the different items included in the index, and allows the items to be added up. Methods to estimate the economic value of social factors (e.g. the cost of unemployment or divorces) are used extensively by social scientists, while ecological economists have developed methods to estimate the economic value of environmental factors (e.g. air pollution). In the following chapter we introduce the GPI, and the thinking behind it, while in Chapter 4 we explain the methods used for each item.

References

Alfsen, K. H., Hass, J. L., Tao, H. and You, W. (2006). International experiences with 'green GDP'. Statistics Norway. Retrieved from www.ssb.no/a/publikasjoner/pdf/rapp_200632/rapp_200632.pdf

Anand, S. and Sen, A. (1997). Concepts or human development and poverty! A multidimensional perspective. *United Nations Development Programme, Poverty and human development: Human development papers*, 1–20.

Brouwer, R. *et al.* (2013). A synthesis of approaches to assess and value ecosystem services in the EU in the context of TEEB. Retrieved from http://ec.europa.eu/environment/nature/biodiversity/economics/pdf/EU%20Valuation.pdf

CBD. (2009). Report of the first expert workshop on the development of the city biodiversity index. Paper presented at first expert workshop on the development of the city biodiversity index, 10–12 February, Singapore City. Retrieved from www.cbd.int/doc/meetings/city/ewdcbi-01/official/ewdcbi-01-03-en.pdf

CBD. (2010a). Report of the second expert workshop on the development of the city biodiversity index. Paper presented at second expert workshop on the development of the city biodiversity index, 1–3 July, Singapore City. Retrieved from www.cbd.int/doc/meetings/city/ewdcbi-02/official/ewdcbi-02-03-en.pdf

CBD. (2010b). User's manual for the city biodiversity index. Retrieved from www.cbd.int/authorities/doc/User's%20Manual-for-the-City-Biodiversity-Index27Sept2010.pdf

Chan, L. (2012). Singapore Index on Cities' Biodiversity. World Cities Summit 2012. Singapore, 3 July 2012. Retrieved from www.worldcitiessummit.com.sg

Chan, Y. K., Kwan, C. C. A. and Shek, T. L. D. (2005). Quality of Life in Hong Kong: The CUHK of Hong Kong Quality of Life Index. *Social Indicators Research*, 71, 259–289. doi: 10.1007/s11205-004-8020-4. Retrieved from www.cuhk.edu.hk/ssc/qol/eventdoc/1214387991e/out.pdf

Codrington, S. (2005). *Planet Geography*. Sydney: Solid Star Press.

Costanza, R., d'Arge, R., de Groot, R., Farber, S., Grasso, M., Hannor, B., Limburg, K., Naeem, S., O'Neill, R., Paruelo, J., Rasins, R., Sutton, P. and van den Belt, M. (1997). The Value of the world's ecosystem services and natural capital. *Nature, 387*, 253–260.

Costanza, R., Hart, M., Talberth, J. and Posner, S. (2009). *Beyond GDP: The need for new measure of progress*. Boston University: The Pardee Papers No. 4, Retrieved from www.bu.edu/pardee/files/documents/PP-004-GDP.pdf

CUHK. (2012a, July 23). CUHK Hong Kong Quality of Life Index: Quality of Life in Hong Kong Declined. Chinese University of Hong Kong. Retrieved 14 October 2014 from www.cpr.cuhk.edu.hk/en/press_detail.php?id=1351

CUHK. (2012b). CUHK Hong Kong Quality of Life Index. Chinese University of Hong Kong. Retrieved from www.cpr.cuhk.edu.hk/resources/press/pdf/500cfb3927731.pdf

CUHK. (2013). CUHK Hong Kong Quality of Life Index: Quality of Life in Hong Kong Rebounded Slightly. Press release 26 September 2013. Retrieved 14 October 2014 from www.cpr.cuhk.edu.hk/en/press_detail.php?1=1&id=1635

EEA. (2007). Environmental pressures from European consumption and production: Insights from environmental accounts. European Environment Agency – Brochure No 1/2007. Downloaded 10 May 2013 from www.eea.europa.eu/publications/ brochure_2007_1/at_download/file

Endl, A. and Sedlacko, M. (2012). National Sustainable Development Strategies – What Future Role with Respect to Green Economy? *European Sustainable Development Network (ESDN)*. Retrieved from www.sd-network.eu/pdf/policy_briefs/ESDN_ UNCSD_Policy_Brief.pdf

Environmental Audit Committee. (2012). Measuring well-being and sustainable development: Sustainable Development Indicators. The House of Commons: London. Retrieved from www.publications.parliament.uk/pa/cm201213/cmselect/cmenvaud/667/667.pdf

Esty, D. C., Levy, M., Srebotnjak, T. and De Sherbinin, A. (2005). Environmental sustainability index: benchmarking national environmental stewardship. *New Haven: Yale Center for Environmental Law & Policy*, 47–60.

Esty, D., Levy, M., Srebotnjak, T., de Sherbinin, A., Kim, K. and Andersson, B. (2006). Pilot 2006 Environmental Performance Index. New Haven: Yale Center for Environmental Law & Policy. Retrieved from www.yale.edu/epi/2006EPI_Report_ Full.pdf

European Commission. (2011). Sustainable development in the European Union 2011 monitoring report of the EU sustainable development strategy. European Union. Retrieved from ec.europa.eu/eurostat/documents/3217494/5731705/224-EN-EN.PDF

Everett, G. and Wilks, A. (1999). The World Bank's Genuine Savings Indicator: a Useful Measure of Sustainability? Bretton Woods Project. Working to reform the World Bank and IMF. 10p. Retrieved from www.brettonwoodsproject.org/topic/environment/ gensavings.pdf

Fiala, N. (2008). Commentary on Measuring sustainability: Why the ecological footprint is bad economics and bad environmental science. *Ecological Economics*, *67*, 519–525. Retrieved from www.colorado.edu/geography/class_homepages/geog_2412_f11/ assignments_discussions/ecological%20footprint%20bad%20science.pdf

Global Footprint Network. (2011). Frequently Asked Technical Questions. Oakland (USA): Global Footprint Network. Retrieved from www.footprintnetwork.org/en/ index.php/GFN/page/frequently_asked_technical_questions/#ic7

Global Footprint Network. (2014). Living Planet Report 2014: Species and spaces, people and places. Oakland (USA): Global Footprint Network.

Goossens, Y. A., Makipaa, A. *et al.* (2007). Alternative Progress Indicators to Gross Domestic Product (GDP) as a Means Towards Sustainable Development. Brussels: European Parliament, Policy Department, Economic and Scientific Policy. Retrieved from http://edz.bib.uni-mannheim.de/daten/edz-ma/ep/07/EST19990.pdf

Hametner, M. and Steurer, R. (2007). Objectives and indicators of sustainable development in Europe: A comparable analysis of European coherence. Brussels: ESDN Quarterly Report.

Havinga, I. (2011). The System of Environmental-Economic Accounting (SEEA) – The measurement and monitoring framework for the environment-economy relationship for official statistics. United Nations Statistics Division Interactive Dialogue of the General Assembly on Harmony with Nature, United Nations. Retrieved from www.uncsd2012. org/content/documents/Ivo%20Havinga.pdf

Hecht, J. E. (2002). Green National Income, Measures of Welfare, And Ideological Bias. Retrieved from www.joyhecht.net/professional/papers/jhecht-Welfare-Measures-Bias-ISEE-Conf-Feb02.pdf

HKCSS. (2002). The Hong Kong Council of Social Service Social Development Index 2002 Major Findings. Hong Kong Council of Social Services. Retrieved from www.hkcss.org.hk/pra/sdi_e.pdf

HKCSS. (2012, March 7). HK Social Development Index 2012. Hong Kong Council of Social Services. Retrieved from www.hkcss.org.hk/pra/Press/SDI_2012_Press_release_5March_eng.pdf

Jackson, T. and McBride, N. (2005). Measuring Progress? A review of 'adjusted' measures of economic welfare in Europe. Prepared for the European Environment Agency, Copenhagen. Retrieved from www.surrey.ac.uk/ces/files/pdf/1105-WP-Measuring-Progress-final.pdf

Jinnan, W., Hongqiang, J. and Fang, Y. (n.d.). Green GDP Accounting in China: Review and Outlook. Chinese Academy for Environmental Planning, Beijing. Retrieved from http://unstats.un.org/unsd/envaccounting/londongroup/meeting9/china_country_report_2004.pdf

Kitzes, J., Galli, A., Bagliani, M., Barrett, J., Dige, G., Ede, S. and Wiedmann, T. (2009). A research agenda for improving national Ecological Footprint accounts. *Ecological Economics, 68*(7), 1991–2007.

Klasen, S. (2006). UNDP's Gender-related measures: Some conceptual problems and possible solutions. *Journal of Human Development, 7*(2), 243–274. Retrieved from www.tandfonline.com/doi/full/10.1080/14649880600768595#preview

Klasen, S. and Schüler, D. (2011). Reforming the Gender Related Development Index and the Gender Empowerment Measure: Implementing Some Specific Proposals, *Feminist Economics, 17*(1), 1–30. Retrieved from www.ccee.edu.uy/ensenian/catgenyeco/Materiales/2011-08-10%20M6%20-%20KlasenShuler(2011).pdf

Krishnaji, N. (1997). Human Poverty Index: A Critique. *Economic and Political Weekly*, 2202–2205.

Land, K. C., Michalos, A. C. and Sirgy, M. J. (2012). *Handbook of Social Indicators and Quality of Life Research*. London: Springer.

Lenzen, M. and Murray, S. A. (2003). The Ecological Footprint – Issue and Trends. ISA Research Paper 01–03. The University of Sydney. Retrieved from www.isa.org.usyd.edu.au/publications/documents/Ecological_Footprint_Issues_and_Trends.pdf

London Group on Environmental Accounting. (25 November 2007). Contribution to Beyond GDP 'Virtual Indicator Expo'. *System of Environmental-Economic Accounting (SEEA)* Paper presented at International Conference, Brussels (1–4). Retrieved from www.beyond-gdp.eu/download/bgdp-ve-seea.pdf

Maro, P. (2007). Ideas for overcoming the limitations of GDP as a progress indicator. *European Environmental Bureau*. Retrieved from www.eeb.org/publication/documents/SDSprogressBriefing-1207.pdf

Morris, D. and McAlpin, M. (1979). *Measuring the Condition of the World's Poor*. New York: Pergamons Press.

NEF. (2012a, June 14). Measuring what matters: the Happy Planet Index 2012. New Economics Foundation. Retrieved from www.neweconomics.org/blog/entry/measuring-what-matters-the-happy-planet-index-2012

NEF. (2012b). Happy Planet Index Report 2012. New Economics Foundations. Retrieved from www.happyplanetindex.org/assets/happy-planet-index-report.pdf

Nordhaus, W. D. and Tobin, J. (1972). 'Is Growth Obsolete?'. *National Bureau of Economic Research*, pp. 1–80. Retrieved from www.nber.org/chapters/c7620.pdf

Parr, S. F. (2006). The human poverty index: A multidimensional measure. Poverty in Focus. *International Poverty Center*. UNDP. Retrieved from www.ipc-undp.org/pub/IPCPovertyInFocus9.pdf

Pinter, L., Hardi, P. and Bartelmus, P. (2005). *Sustainable development indicators: proposals for the way forward prepared for the United Nations Division for Sustainable Development*. International Institute for Sustainable Development.

Rodricks, S. (2010). Singapore City Biodiversity Index. Retrieved from www.teebweb.org/wp-content/uploads/2013/01/Singapore-city-biodiversity-index.pdf

Sen, A. (2004). Capabilities, lists, and public reason: continuing the conversation. *Feminist Economics*, *10*(3), 77–80.

Stewart, K. (1974). National Income Accounting and Economic Welfare: The Concepts of GNP and MEW. *Federal Reserve Bank of St. Louis*. Retrieved from http://research.stlouisfed.org/publications/review/74/04/Accounting_Apr1974.pdf

UN. (2003). Handbook of National Accounting. Integrated Environmental and Economic Accounting 2003. Final draft circulated for information prior to official editing. Studies in Methods, Series F, No.61, Rev.1 (ST/ESA/STAT/SER.F/61/Rev.1). Geneva: United Nations http://unstats.un.org/unsd/envaccounting/seea2003.pdf

UNDP. (1995). *Human Development Reports*. Geneva: United Nations Development Programme. Retrieved from http://hdr.undp.org/en/humandev/

UNDP. (2006). Human Development Report. (2006). Beyond scarcity: Power, poverty and the global water crisis. United Nations Development programme. Retrieved from http://hdr.undp.org/en/media/HDR06-complete.pdf

UNDP. (2012). Millennium Development Goals. Geneva: United Nations Development Programme. Retrieved from www.un.org/millenniumgoals/reports.shtml

Wackernagel, M. and Rees, W. (1996). *Our Ecological Footprint: Reducing Human Impact on the Earth*. Gabriola Island: New Society Publishers.

Wackernagel *et al.* (2002). Tracking the ecological overshoot of the human economy. *Proceedings of the National Academy of Sciences*, *99*(14), 9266–9271. Retrieved from www.ncbi.nlm.nih.gov/pmc/articles/PMC123129/pdf/pq1402009266.pdf

Wu, J. and Wu, T. (2010). Green GDP. *Berkshire Encyclopedia of Sustainability*, *2*, 248–250. Retrieved from http://leml.asu.edu/jingle/Web_Pages/Wu_Pubs/PDF_Files/Wu+Wu-2010-GreenGDP.pdf

World Bank. (2005). Where is the wealth of nations? The International Bank for Reconstruction and Development/The World Bank. Retrieved from http://siteresources.worldbank.org/INTEEI/214578- 1110886258964/20748034/All.pdf

World Bank Staff. (1997). *World development report 1997: the state in a changing world*. Oxford: Oxford University Press.

WWF. (2012). Living Planet Report 2012. Retrieved from www.wwf.de/fileadmin/fm-wwf/Publikationen-PDF/WWF_LPR_2012.pdf

Xu, X. (2011). Methodology of GDP estimation in the year of the first economic census of China. *Department of National Accounts, NBS*. Retrieved on 25 November 2011 from: www.oecd.org/dataoecd/45/13/36561504.pdf.

3 The Genuine Progress Indicator as an alternative indicator of welfare

Introduction

In the early 1970s, economists began to challenge the assumption that economic growth is proportionate to improvements in human welfare (Berik and Gaddis, 2011). Some suggested that emphasis should be placed on optimal scale, fair distribution, improved institutional arrangements, efficient allocation of resources, and qualitative rather than quantitative growth, in order to achieve sustainable development (Clarke and Shaw, 2008; Berik and Gaddis, 2011). The Genuine Progress Indicator is a step towards assessing the extent to which these goals have been realized.

The Genuine Progress Indicator (GPI) was developed in 1994 by Clifford Cobb, Ted Halstead and Jonathan Rowe of Redefining Progress (Cobb *et al.*, 1995). The idea behind the GPI was that of accounting for the different elements which accompany economic growth and affect human welfare, or people's quality of life. Most studies divide those elements into three different major categories – social, economic and environmental, which is consistent with the three pillars of sustainability. As such, the GPI is calculated by adding the monetary value of the economic, social and environmental goods, resources or services that improve the quality of life, and subtracting the monetary value of those that worsen the quality of life. Additional categories, for example political and spiritual, have also been used (e.g. Clarke and Shaw, 2008).

Chapter 4 provides the complete list of all the items included in the GPI, as well as a description of the methodology used to estimate their economic value. The positive items used to estimate the GPI include: 1) personal consumption (where defensive/rehabilitative expenditure are subtracted); 2) service from publicly provided infrastructure; 3) value of non-paid household labour; and 4) value of volunteer labour. The negative items subtracted from the GPI, which are either ignored or added to the GDP, include: 1) unemployment and underemployment; 2) crime; 3) overwork/loss of leisure time; 4) non-renewable resource depletion; 5) environmental degradation; and 6) harm from air, water and noise pollution.

Like the GDP, the GPI measures the welfare of an area over a period of time, which allows one to identify trends and illustrate the source of positive and negative changes (Hamilton, 1999; Berik and Gaddis, 2011). However, unlike the GDP, the GPI also takes into consideration the social and environmental costs and benefits that accompany economic growth. As such, the GPI addresses some of the problems that exist with the calculations of the GDP, and provides a better estimate of the level of welfare or well-being of its citizens. Furthermore, the GPI helps policymakers identify weaknesses, increases public awareness, and is easy to understand due to its structural similarities to the GDP (Goossens *et al.*, 2007). This chapter introduces the GPI, reviews its theoretical justifications, and discusses its advantages and disadvantages.

The concepts of weak and strong sustainability

The concept of sustainable development was given impetus at the 1972 Stockholm Conference, which produced a set of principles that promoted concepts of intergenerational equity. The Brundtland Commission came up with the most well-known definition of sustainable development: 'Sustainable development is development that meets the needs of the present without compromising the ability of future generations to meet their own needs' (Brundtland, 1987). However, only at the 1992 Rio Conference was the concept of ecologically sustainable development made explicit, with emphasis on the precautionary principle, intergenerational equity, the conservation of biological diversity and ecological integrity, and improved valuation, pricing and incentive mechanisms (Bates, 2010). This concept was especially important in persuading people to accept the need to keep stocks of man-made capital and natural capital at levels that could maintain a sustainable rate of production and consumption without jeopardizing current and future well-being (Berik and Gaddis, 2011).

Sustainability depends on the relationship between two types of capital: man-made capital and natural capital. Man-made capital includes all goods and services provided through man-made production facilities (physical machinery, factories, buildings and infrastructure). Because this capital tends to depreciate in value, a portion of income needs to be set aside to maintain their productivity (Hamilton, 1999; Anielski, 2001). By contrast, natural capital comprises the means to sustain future production and consumption through the supply and extraction of low-entropy resources (i.e. basic and unprocessed resources, whether renewable or non-renewable, such as timber, minerals and fossil fuels). Also inherent in natural capital are natural environmental services (such as the protective ozone layer), and nature's capacity to assimilate or dissipate high-entropy waste products and prevent them from harming humans and other species (Hamilton, 1999). Natural capital, like man-made capital, depreciates in both physical and market value terms, and therefore requires continued investment to maintain productivity and ongoing throughput of energy matter (Anielski, 2001).

Sustainability relies on maintaining the productivity of these stocks of capital, which provide for the level of production and consumption we require. This raises the question of whether man-made capital and natural capital should be kept separate, or whether man-made capital would be an adequate substitute for natural capital (Lawn, 2003; Berik and Gaddis, 2011). The issue of substitutability of natural capital by man-made capital has been dealt with by the 'weak sustainability' versus 'strong sustainability' debate.

Weak sustainability allows for the substitution between human capital and natural capital, which means that a decline in natural capital (e.g. a destruction of a wetland) may be partly compensated by an increase in human capital (e.g. the construction of a sewage treatment plant which replaces some functions of the wetland). By contrast, strong sustainability suggests that man-made capital is not an adequate substitute for natural capital. Strong sustainability assumes that human capital and natural capital are complementary, but not interchangeable. Strong sustainability regards the natural environment as the sole source of low-entropy resources, the sole repository and assimilator of high-entropy waste, and critical to life-support services such as clean air and water. From the point of view of strong sustainability it is crucial to maintain both capital stocks if sustainable well-being is to be achieved (Lawn, 2003).

Some have argued that technological advances embodied in man-made capital should reduce the need for natural capital to fuel further economic progress. However, Lawn (2003) indicated that this would not equate to substitution, for three reasons.

First, technological advances in economic processes merely reduce the high-entropy waste generated from the process of transforming natural capital to man-made capital. As a result of the first and second laws of thermodynamics – no 100 per cent production efficiency, no 100 per cent recycling of materials, and the impossibility of recycling energy – there will always be a need for the throughput of energy matter, and therefore the need for low-entropy resources provided only from natural capital.

Second, in line with the first reason, the elasticity of substitution (i.e. how easy it is to substitute one input for the other) between man-made capital and natural capital is less than one – where a value of at least one is required to demonstrate adequate substitutability – and it 'tends toward zero as attempts are made to augment man-made capital to offset the impact of declining natural capital' (Lawn, 2003: 110).

Third, the quantity of natural capital needed to provide critical life-support services would far exceed the quantity needed to maintain economic processes.

For these reasons, strong sustainability is usually considered a better approach to assess the level of sustainability of a country. When using the strong sustainability approach to measure sustainable incomes, the estimated depletion value of natural capital must reflect the cost required to keep natural capital stocks intact. This would involve setting aside a portion of the proceeds from the exploitation of the resources, to be invested into the cultivation of additional stocks of, or substitutes for, renewable resources. The 'user cost' formula devised by El Serafy (1989) (see Chapter 4) allows for the calculation of the set-aside amount and the replacement/substitute cost of resource depletion, leaving the remainder as legitimate income. To calculate national incomes in accordance with the strong sustainability concept, the user cost must be subtracted from the GDP to approximate the investment costs needed to keep natural capital intact (Lawn, 2003).

The Genuine Progress Indicator (GPI) does effectively illustrate the use of different forms of capital and their physical conditions (Anielski, 2001). However, it has been argued that the final aggregation of values in the GPI should not be taken as a measure of strong sustainability (Wen *et al.*, 2007). This is because the GPI reports the aggregation of different welfare indicators. If man-made capital increases and natural capital decreases, the overall GPI may remain the same, suggesting the substitutability of natural capital. It may be argued that the GPI does incorporate the concept of intergenerational equity in its calculations, by accounting both for past costs and ongoing costs (as discussed in Chapter 4). However, the GPI alone does not determine whether past and present economic activities and the benefits they bring, are more or less sustainable (Wen *et al.*, 2007; Lawn and Clarke, 2010).

Because of these considerations, it would be helpful to supplement GPI figures with those of other indices, such as Ecological Footprint, to determine whether the welfare is sustainable. The Ecological Footprint assesses human demand on the Earth's ecosystems, comparing the rate of exploitation with the rate of regeneration and calculating how far human consumption and production have exceeded the planet's carrying capacity (Wen *et al.*, 2007; Lawn and Clarke, 2010), in other words, whether economic processes comply with the concept of strong sustainability.

The concepts of welfare and wealth, income and capital

The concepts of welfare and wealth can mean different things to different people, and since the GPI is a measure of welfare, it is important to understand what exactly we mean. To Anielski (2001), the fundamental definition of wealth is based on the Old English word '*weal*', which means 'the condition of well-being'. The word 'economy', originating from the Greek word '*oikonomia*', is defined as 'the management of the household' (Anielski, 2001). Therefore, in principle, the economy and the accumulation of wealth should focus on improving the well-being of households, as well as the conditions of the natural environment that contribute to human well-being (Anielski, 2001). Current definitions of wealth and economy largely understate this fundamental principle, and therefore wrongly represent economic production (as measured by the GDP) as indicative of quality of life and well-being, while they fail to account for problems of economic instability, income inequality, social disorder and environmental degradation.

Hicksian and Fisherian income and capital

The GPI takes consumer expenditure as the foundation of welfare. For this reason, besides the concepts of welfare and wealth, the definitions of income and capital are critical in the study of the GPI. Two interpretations of income and capital are important in understanding the concept of the GPI: that of John

Hicks (1946) and that of Irving Fisher (1906). Hicks described income as the maximum amount a person or a nation can consume over a period of time, and still be as 'well off' at the end of that period as at the beginning (Hicks, 1946: 172, in Hamilton, 1999). Hence, income is the maximum sustainable consumption: 'Sustaining consumption over a given period depends on maintaining the productive potential of the capital stocks that are needed to generate the flow of goods and services that are consumed' (Hamilton, 1999: 7). This concept is consistent with the aforementioned definition of wealth and economy. Hicksian income requires the constant flow of services provided by capital stocks and, therefore, requires that a portion of current consumption be 'set aside' to replenish stocks affected by depreciation or depletion (Hamilton, 1999; Lawn and Clarke, 2008b). The GPI attempts to estimate the national income in Hicksian terms, since it subtracts from the GDP the depreciation of man-made capital, the depletion of natural capital, and the defensive and rehabilitative expenditures (some of which are necessary to maintain the capital stock). The resultant sum is the maximum amount a nation can consume while continuing to be as well off in the future as it is today. In spite of its usefulness, Hicksian's theory of income has several limitations.

First, it suggests that welfare is directly associated with levels of production and consumption, and ignores other factors that are at least as important, such as crime, family breakdown, volunteer and non-paid labour. In contrast, Fisher incorporates the concept of psychic enjoyment of services, including considerations of the quality of the stock and income distribution, which are critical to overall welfare. Second, Hicks' definition does not take into account properly the timeframe between the investment in capital or consumer goods, and the period during which the goods are consumed. Since capital goods and consumer goods are consumed for a period of time, the benefits of such consumption should be accounted for and distributed throughout the lifetime of the product, rather than on the time of purchase. Third, Hicks' concept does not account for the depreciation of consumer durables.

On the other hand, Fisher's definition of income focuses not on the goods produced and consumed in a particular year, but on the services enjoyed by the ultimate consumers of all man-made capital, or the 'subjective utility yielded by goods and services consumed' (Li, 2006: 45). This he called 'psychic income' or 'utility satisfaction' (Lawn, 2003). This concept means that expenditure on durable goods is accounted for as additions to the stock of man-made capital, and their benefits are spread over the lifetime of the products, rather than only during the year the expenditures were made, as is done with the GDP (Lawn, 2003). Capital is defined as goods subjected to human ownership and capable of providing the direct or indirect satisfaction of human needs and wants. It therefore encompasses both producer and consumer goods (Lawn, 2003). This concept of income and capital also emphasizes the importance of setting aside a portion of current consumption to pay for maintaining the man-made and natural capital that are necessary for the production of new goods (Lawn, 2003). According to Lawn (2003), Lawn and Clarke (2008b; 2008c)

Fisher's concept of income and capital best reflects the theory behind, and methods used by, the GPI.

Indeed, the concept of welfare in the GPI involves, first, the concept of the psychic benefits (or psychic income) a consumer enjoys from a purchased product (or, more broadly, from natural capital, human capital or social capital), whether long-term or short-term. This idea was introduced by Irving Fisher (1906) and related to the GPI by Philip Lawn (2003), who defines welfare as the services that are enjoyed by the ultimate consumers of man-made goods.

Consumption expenditure as the basis of GPI

To better understand the reason for basing the GPI on consumption expenditure, it is useful to compare the way the GPI is calculated to the way in which the GDP is calculated. The GDP is calculated as the sum of private-sector consumption and investment expenditure (i.e. expenditure on man-made capital), public-sector consumption and investment expenditure, and exports, while the value of imports is subtracted. On the other hand, since the GPI aims to calculate the welfare generated directly from domestic economic activity, exports are excluded while imports are retained, private and public expenditure are brought together as consumption expenditure (serving as the foundation of GPI), and the benefits of investments are spread over the years people enjoy their benefits, according to Fisher's definition.

Strength and weaknesses of the GPI

The GPI has a number of strengths and weaknesses, both at the conceptual and at the methodological levels. Some of these strengths and weaknesses are now discussed.

Strengths of the GPI

1. It is an holistic indicator that stresses the interrelationship between different complex variables

The GPI uses a number of different economic, social and environmental items to estimate a population's welfare. These items aim to reflect not only the size of the economic output (as does the GDP), but also the costs that (may) accompany that output, in terms, for example, of additional environmental degradation and social ills. An indicator of people's welfare or standard of living should also include a diversity of psychological and emotional conditions, such as happiness, hope, fear, etc. Unfortunately, these are difficult to quantify, and the GPI ignores them. Nevertheless, the GPI provides a more complete picture of a population's welfare, and can therefore reveal weaknesses or foster policies that can improve our sustainable well-being (Anielski, 2001; Clarke and Lawn, 2008). In other words, the GPI is able to provide an insight into whether the current economic path is sustainable, and how a more sustainable economy can be achieved (Clarke and Lawn, 2008).

2. It provides an easily understood indicator

The GPI retains its focus on the economic aspects. Also, since the non-market costs and benefits are given a monetary value, the GPI is presented as a monetary value (e.g. in HK$ or Singapore $), in the same way as the GDP. As such, it can also be more easily understood by policymakers and the general public, can be compared to the GDP, and can more easily replace the GDP in public imagination and official discourse

3. It allows for historical analysis and comparison among countries

While a large number of data are required to compile the GPI, in many countries these data are already collected, which would facilitate its widespread use, and allow for comparison between countries. Furthermore, in many cases these data have been collected for a number of years, which means that the GPI can be estimated back in time. This is a clear advantage when compared to other indices described in Chapter 2 (e.g. the HKQLI and the SDI), which cannot be estimated for past years. GPI studies usually span over long periods (a few decades), and reveal trends that may help us understand the long-term consequences of certain policies, or long-term changes in welfare (see Chapter 7). In the present case, we estimate the GPI of Hong Kong and Singapore from 1968 to 2010.

4. It is methodologically strong

All economic, social, and environmental items of the GPI are given an economic value. This allows items to be compared, added and subtracted, so as to arrive at a value that represents the overall GPI of a particular year. Giving an economic value does not mean that everything is (or should be) commodified. Economic values are simply used as a common unit that allows unrelated variables (e.g. air pollution and family breakdown) to be summed up. Economic valuation is used as a benchmark, because we are familiar with the concept of money, not because we aim to commodify all social and environmental variables.

At the same time, by giving an economic value to the items, they are given a weight which corresponds to people's value.[1] This is the case both for those items that are based on people's expenditures or investments (such as *Defensive and rehabilitative expenditures*), and for those that are estimated using government expenditure to address environmental problems (e.g. *Cost of waste-water pollution*). As long as the political process works effectively (i.e. public investments are made according to people's wishes), government expenditure corresponds to the value that society puts on the environmental degradation. As such, we believe that the GPI addresses some of the weaknesses of the other indices, and can be considered a conceptually sound approach (Chapman, 1999; Field and Field, 2005; Tietenberg, 2005; Field, 2008). However, one weakness of such approach is that only the contribution to human welfare is included. For example, only the costs of climate change to society are included, not the intrinsic value of species that may become extinct because of climate change.

5. It helps determine the threshold and call for a steady state economy

The GPI informs policymakers as to the need to build a sustainable economy. Using the GDP as the indicator of a country's development encourages the pursuit of endless economic growth, since the GDP does not account for the negative consequences that economic growth has on environmental and social systems. On the other hand, the GPI accounts for the environmental and social costs, and informs us when these costs outstrip the benefits of economic growth. Indeed, assessing a country's development using the GPI has shown that when countries reach a certain level of economic development, there is a 'threshold' beyond which further economic development no longer increases welfare (Max-Neef, 1995) (discussed in Chapter 7). At that point, further economic growth lowers welfare and increases environmental degradation, and a 'steady state economy' becomes more desirable than a 'growth economy'. The steady state is an economy that is geared towards a dynamic equilibrium with the ecosystem that supports it, by means of qualitative improvement of existing goods and capital, instead of an increase in the production of goods and the consumption of stocks, which the growth economy depends on. It emphasizes the quality of goods and services over their quantity, a concept which embodies a sustainable use of natural resources and a more equitable distribution of income among citizens. This, and the policies necessary to make the transition, will be discussed in Chapter 8.

Weaknesses of the GPI

There have also been a number of criticisms levelled at the GPI (see Brennan, 2008; Harris, 2007; Neumayer, 1999, 2000 for further elaboration).

1. Lack of standardization in the choice of items

The GPI has been criticized for inconsistency among countries, in terms of the items used. That is, it is not (yet) a standardized measure, and researchers include different items when estimating the GPI of different countries. We contend that this is not necessarily a problem. The addition of particular items could be seen positively, as an attempt to better reflect the conditions of each country under investigation. Indeed, one may argue that it is quite logical to have different sets of indices to accurately reflect the welfare of different countries, since no two countries are the same. For example, the indicator *Loss and damage to terrestrial ecosystems* was included in the New Zealand GPI because of the serious harm caused by invasive pests, which had been identified as the greatest threat to biodiversity (Forgie *et al.*, 2008). Similarly, the study of Thailand (Clarke and Shaw, 2008) included the cost of corruption, since corruption continues to be severe, and damage the development of the country. The choice of the variables, and the weight given to them, really depends on how each of these variables are defined and justified in the national context.

The question also arises as to how to address changing social norms. For example, an increase in family breakdown might be indicative of a greater social acceptance of divorce (especially in conservative societies such as Hong Kong and Singapore) rather than an indicator of social upheaval. Similarly, changing values may gradually weaken the social importance of non-paid (female) household labour, even in a Chinese society that aims at following Confucian principles. It is difficult to decide how to deal with these gradual changes in social values, and to adapt these changes in a quantitative index. In this study, to facilitate comparison, we have tried to use the same items and method used by others, but one may as well argue for the use of different items.

At the same time, particular items may be missing for some countries due to the absence of data. For example, Lawn (2008b) excludes the cost of underemployment due to data limitations (for India), and Berik and Gaddis (2011) extrapolate net capital investment and the value of volunteer labour (for the US state of Utah). If we expected all studies to include the same indicators, we should exclude all indicators that are not available for any country, settling for the lowest common denominator.

The issue of inconsistency, and the potential difficulty in comparing results among countries, could be an area of concern because comparisons help us extrapolate results. Nevertheless, we can argue that an index is of greater use to a country if it accurately represents its conditions, rather than if it is directly comparable to other countries. The items chosen to estimate the GPI should then be able to represent the development of a country's sustainable welfare. As such, there are a number of 'base' indices (discussed in Chapter 4) which guarantee some measure of uniformity, to which other items are included to reflect as accurately as possible the conditions of the countries under investigation. The issue perhaps is not that of standardizing the items that are included, but that of standardizing the methods by which they are calculated.

We contend that there may not be a need for a standard list of items, and we may as well use additional ones as long as: 1) the inclusion of such items is justified in the national context; 2) these items allow us to better understand the conditions of the country under investigation; 3) they are methodologically consistent (for example they do not involve double-counting); and 4) the additional items do not transform the GPI so much that comparisons with other countries is impossible.

2. Difficulties in giving an economic value to non-marketed goods and services

Giving a monetary value to non-marketed goods, whether environmental or social, is difficult since they do not have a clear price. The problem is particularly important with externalities. Externalities are third party (or spill-over) effects of economic activities (usually arising from the production and/or consumption of goods) whose costs (or benefits) are not included in the market price. Examples include air pollution as a byproduct of electricity production, whose costs (in terms of hospital visit and reduced visibility for example) are not included in the price of the electricity, but instead are borne by the people. Valuing these costs is difficult because costs may vary (they may range from a general feeling of discomfort, to bronchitis which may cause losses in work days, or to lung cancer), and it is usually difficult to pinpoint the exact source of the externality. Ecological economists have developed methods to estimate the value of externalities. However, they are not fool-proof, and are open to criticisms.

Estimating other environmental costs is also problematic, and open to challenge. For example, the value of some environmental degradation is estimated using government expenditure to clean up that degradation. However, if the government is not overly concerned with environmental degradation, and spends little money to clean it up, then the GPI records little environmental degradation. Similarly, in the GPI the cost of fishery depletion is calculated based at the catch. This means that if all fish are caught and the catch drops to zero, the GPI reports the cost of fishery depletion to be 0.

Socio-economic variables, such as the costs of divorce, volunteer labour and household labour are almost as difficult to estimate. One of the difficulties is to estimate the value of time. Different approaches exist (e.g. the minimum wage rate, or the average rage rate, or a proportion of them) and can be justified. Chapter 4 explains in detail the methods we use for each of the environmental and social items we include in the GPI. Whenever possible, we have chosen to replicate the approaches used by others, so as to facilitate comparison among studies.

3. It is an indicator of weak sustainability

As mentioned above, the concept of weak sustainability states that man-made capital can substitute natural capital, while the principle of strong sustainability states that natural capital cannot be substituted. The GPI includes both man-made capital (economic and social variables), and natural capital (environmental variables). The final result from the calculation of the GPI is the sum of all positive and negative values from all three (economic, social and environmental) variables, and therefore implies the substitutability of natural capital. Because of this, the GPI has been criticized for being an indicator of weak sustainability, and its ability to accurately reflect the sustainability of a country has been challenged. However, if we are seeking an index that is able to capture the three pillars of sustainability and display them in an easy-to-understand manner (for example with monetary values), it is inevitable that such index is an indicator of weak sustainability. The GPI is undoubtedly superior to the GDP when it comes to presenting the genuine progress of a country (Lawn and Clarke, 2008e).

4. It is not adapted to the aspirations of different groups

The definition of well-being varies among people because people have different needs and aspirations, which are shaped by culture, age, ethnicity, economic background and more. For this reason, it is debatable how far the GPI can be a comprehensive indicator of well-being. Similarly, if the GPI is a measure of welfare, it would be useful to understand how different groups (such as the poor, single-parent families, elderly, migrants, minorities) have been performing over time, and how conditions have changed. Unfortunately, this is hardly possible with the GPI, as the datasets available are at the national level. Nevertheless, the GPI does include an *Income distribution index* (estimated through the GINI coefficient) to give some weight to income inequality (see Chapter 4).

This, of course, is a drawback, since many government policies target specific groups, and an indicator that is able to determine the well-being of particular groups, apart from the population as a whole, would be very useful. Other indicators, such as the Social Development Index (SDI, introduced in Chapter 3) in the case of Hong Kong, would help in assessing the standard of living of particular groups. Perhaps a combination of different indicators, each using a different method and each including different variables, would be a better approach to gauge social well-being.

5. Scale

A strong argument can be made that the pollution generated – and natural resources consumed – for the production of goods to be exported should be included among the data of the importing country, rather than the exporting country. This is particularly so for city-states, since the overwhelming majority of industrial goods and food products (which generate great amounts of pollution) are imported. Unfortunately, there are not yet sufficiently comprehensive datasets that can be used to address this problem.

Conclusions

This chapter has reviewed the history of the GPI, the thinking underpinning it, its justifications, and its advantages and disadvantages. We believe that the GPI is a good indicator of the welfare of a country, since it includes a relatively large number of variables, and it attempts to be methodologically sound, bringing together all variables through the use of a common yardstick: money. This does not mean that everything should, or can, have an economic value. Money is used because people are more familiar with money than with any other property or commodity that could be used as a yardstick. This makes the index more easily calculable, and more understandable by people, news networks, and policymakers. Furthermore, using money makes it easy to compare the GPI to the GDP. This does not mean that the GPI does not have weaknesses. We discussed some issues with the way in which the GPI is calculated. We believe that such issues should not detract from its usefulness. After all, the accuracy of official government data is also open to debate, but these data are nevertheless used in government planning (Webster, 2002; Handbury *et al.*, 2013). The primary purpose of the GPI is to contribute to the debate as to 'what sustainable welfare is and how societies might best maintain or increase it' (Clarke and Lawn, 2008: 573), and we believe it does that just fine.

Note

1 We mentioned in Chapter 2 that one weakness of some indices was the lack of weighing the different items included.

References

Anielski, M. (2001). Measuring the Sustainability of Nations: The Genuine Progress Indicator system of Sustainable Well-being Accounts. The Fourth Biennial Conference of the Canadian Society for Ecological Economics: Ecological Sustainability of the Global Market Place, Montreal.

Bates, G. (2010). *Environmental Law in Australia* (7th edn). Australia: LexisNexis.

Berik, G. and Gaddis, E. (2011). The Utah Genuine Progress Indicator, 1990 to 2007: A Report to the People of Utah. Available online: www.utahpop.org/gpi.html

Brennan, A. J. (2008). Theoretical foundations of sustainable economic welfare indicators – ISEW and political economy of the disembedded system. *Ecological Economics*, 67(1), 1–19.

Brundtland, H. (1987). *Our Common Future.* Oxford: Oxford University Press for the World Commission on Environment and Development.

Chapman, D. (1999). *Environmental Economics: Theory, application and policy.* Hong Kong: Addison Wesley.

Clarke, M. and Lawn, P. (2008). Is measuring genuine progress at the sub-national level useful? *Ecological Indicators*, 8, 573–581.

Clarke, M. and Shaw, J. (2008). Genuine progress in Thailand: a systems-analysis approach. In: Lawn, P. and Clarke, M. (Eds), *Sustainable Welfare in the Asia-Pacific: Studies using the Genuine Progress Indicator.* Cheltenham: Edward Elgar Publishing Ltd, 260–298.

Cobb, C., Halstead, T. and Rowe, J. (1995). If the GDP is up, why is America down?. *ATLANTIC-BOSTON*, 276, 59–79.

El Serafy, S. (1989). The proper calculation of income from depletable natural resources. In: Ahmad, Y., El Serafy, S. and Lutz, E. (Eds). Environmental Accounting for Sustainable Development. World Bank, Washington DC: 10–18.

Field, B. C. (2008). *Natural Resource Economics An Introduction* (2nd edn). Long Grove (Ill.): Waveland Press.

Field, B. C. and Field, M. K. (2005). *Environmental Economics* (4th edn). Hong Kong: McGraw-Hill.

Fisher, I. (1906). *Nature of Capital and Income.* New York: A.M. Kelly.

Forgie, V., McDonald, G., Zhang, Y., Patterson, M. and Hardy, D. (2008). Calculating the New Zealand Genuine Progress Indicator. In: Lawn, P. and Clarke, M. (Eds), *Sustainable Welfare in the Asia-Pacific: Studies using the Genuine Progress Indicator* (pp. 126–152). Edward Elgar Publishing Ltd, Cheltenham.

Goossens, Y. A., Makipaa, A., *et al.* (2007). Alternative Progress Indicators to Gross Domestic Product (GDP) as a Means Towards Sustainable Development. Brussels: European Parliament, Policy Department, Economic and Scientific Policy. Retrieved from http://edz.bib.uni-mannheim.de/daten/edz-ma/ep/07/EST19990.pdf

Hamilton, C. (1999). The genuine progress indicator methodological development and results from Australia. *Ecological Economics, 30*, 13–28.

Handbury, J., Watanabe, T. and Weinstein, D. E. (2013). *How Much Do Official Price Indexes Tell Us About Inflation?* (No. w19504). National Bureau of Economic Research.

Harris, M. (2007). On income, sustainability and the 'microfoundations' of the Genuine Progress Indicator. *International Journal of Environment, Workplace and Employment, 3*(2), 119–131.

Hicks, J. (1946). *Value and Capital* (2nd edn). London: Clarendon.

Lawn, P. (2003). A theoretical foundation to support the Index of Sustainable Economic Welfare, Genuine Progress Indicator, and other related indexes. *Ecological Economics, 44*, 105–118.

Lawn, P. (2004). To operate sustainably or not to operate sustainability? – That is the long-run question. *Futures, 36*, 1–22.

Lawn, P. (2005). An assessment of the valuation methods used to calculate the Index of Sustainable Economic Welfare (ISEW), Genuine Progress Indicator (GPI), and Sustainable Net Benefit Index (SNBI). *Environment, Development and Sustainability, 7*, 185–208.

Lawn, P. (2006). An assessment of alternative measures of sustainable economic welfare. In: P. Lawn (Ed.), *Sustainable development indicators in ecological economics* (Chapter 7, pp. 139–165). Northampton, MA: Edward Elgar.

Lawn, P. (2008a). Genuine progress in Australia: time to rethink the growth objective. In: Lawn, P. and Clarke, M. (Eds), *Sustainable welfare in the Asia-Pacific: studies using the genuine progress indicator* (pp. 91–125). Cheltenham: Edward Elgar Publishing Ltd.

Lawn, P. (2008b). Genuine progress in India: some further growth needed in the immediate future but population stabilization needed immediately. In: Lawn, P. and Clarke, M. (Eds), *Sustainable welfare in the Asia-Pacific: studies using the genuine progress indicator* (191–227). Cheltenham: Edward Elgar Publishing Ltd.

Lawn, P. (2010). Facilitating the transition to a steady-state economy: Some macroeconomic fundamentals. *Ecological Economics, 69*, 931–936.

Lawn, P. and Clarke, M. (2008a). An Introduction to the Asia-Pacific region. In: Lawn, P. and Clarke, M. (Eds), *Sustainable welfare in the Asia-Pacific: studies using the genuine progress indicator* (pp. 3–34), Cheltenham: Edward Elgar Publishing Ltd.

Lawn, P. and Clarke, M. (2008b). Why is Gross Domestic Product an inadequate indicator of sustainable welfare. In: Lawn, P. and Clarke, M. (Eds), *Sustainable welfare in the Asia-Pacific: studies using the genuine progress* indicator (pp. 35–46), Cheltenham: Edward Elgar Publishing Ltd.

Lawn, P. and Clarke, M. (2008c). What is the Genuine Progress Indicator and how is it typically calculated. In: Lawn, P. and Clarke, M. (Eds), *Sustainable welfare in the Asia-Pacific: studies using the genuine progress* indicator (pp. 47–68), Cheltenham: Edward Elgar Publishing Ltd.

Lawn, P. and Clarke, M. (2008d). In defence of the Genuine Progress Indicator. In: Lawn, P. and Clarke, M. (Eds), *Sustainable welfare in the Asia-Pacific: studies using the genuine progress indicator* (pp. 69–88). Cheltenham: Edward Elgar Publishing Ltd.

Lawn, P. and Clarke, M. (2008e). Genuine progress across the Asia-Pacific region: comparisons, trends, and policy implications. In: Lawn, P. and Clarke, M. (Eds), *Sustainable welfare in the Asia-Pacific: studies using the genuine progress indicator* (pp. 333–361). Cheltenham: Edward Elgar Publishing Ltd.

Lawn, P. and Clarke, M. (2010). The end of economic growth? A contracting threshold hypothesis. *Ecological Economics, 69,* 2213–2223.

Li, C. Z. (2006). Fisher, Lindhal and Hicks on Income: A modern analysis. In Aronsson, T., Axelsson, R. and Brännlund, R. (Eds), *The theory and practice of environmental and resource economics* (Chapter 3). Cheltenham: Edward Elgar.

Max-Neef, M. (1995). Economic growth and quality of life: a threshold hypothesis. *Ecological Economics, 15*(2), 115–118.

Neumayer, E. (1999). The ISEW – Not an index of sustainable economic welfare. *Social Indicators Research, 48,* 77–101.

Neumayer, E. (2000). On the methodology of the ISEW, GPI, and related measures: Some constructive suggestions and some doubt on the threshold hypothesis. *Ecological Economics, 34,* 347–361.

Tietenberg, T. (2005). *Environmental and Natural Resource Economics* (7th edn). Hong Kong: Addison Wesley.

Webster, D. (2002). Unemployment: how official statistics distort analysis and policy, and why. *Radical statistics, 79,* 96–127.

Wen, Z., Zhang, K., Du, B., Li, Y. and Li, W. (2007). Case study on the use of genuine progress indicator to measure urban economic welfare in China. *Ecological Economics, 63,* 463–475.

4 Items used to calculate the Genuine Progress Indicator

The GPI is calculated using several basic items that are related to the welfare of a country's population. The GPI is not yet a standardized measure, so different studies may use different items. Using the same items has the advantage of facilitating comparisons among countries. However, items may be added depending on the availability of data, since more often than not the GPI uses official, nationwide figures, and countries provide different statistics. In addition, as discussed in Chapter 3, the GPI can, and should, be adapted to the situation of each country, so items may be added to reflect national characteristics. In this study of Hong Kong and Singapore we try to retain the same items and methodologies as other studies. Table 4.1 lists the indices used in the calculation of the Hong Kong and Singapore GPI. The sign that follows the indices shows whether they are added or deducted from the GPI.

Not all items listed in Table 4.1 are included in both Hong Kong's and Singapore's GPI because some items are missing in either Hong Kong or Singapore, or are organized or named differently. For example, Hong Kong has 'Cost of crime', while Singapore has 'Cost of security and external relations', which also includes its expensive armed forces. These differences are only found among the Social and Environmental items. The Economic items are the same, proof of the worldwide emphasis given to the standardization of economic indicators, but relative neglect of environmental and social indicators. The Hong Kong GPI includes 24 indices, and the Singaporean one includes 19 different indices, some of which merge different Hong Kong indices (for example, the Singaporean *Cost of environmental degradation* includes the Hong Kong indices *Cost of waste-water pollution*, *Cost of noise pollution* and *Cost of solid waste*). Overall, we believe that the differences are slight, and we attempt to reduce them further in the calculations.

In the following pages we describe the rationale of the choice of items used in Hong Kong and Singapore, review the methods used by other studies, and describe the methodology we employ. In Chapter 5 we estimate the GPI of Hong Kong, and in Chapter 6 that of Singapore. Our data (unless otherwise specified) come from the 'Hong Kong Annual Digest of Statistics' of the Census and Statistics Department of the Government of the Hong Kong Special Administrative Region (SAR), and from the 'Yearbooks of Statistics' of the Department of Statistics Singapore (DOS). All data in Chapters 5 and 6 are adjusted for inflation.

Table 4.1 Items used for the calculation of Hong Kong's and Singapore's GPI

Economic		Social	Environment
Weighted adjusted consumption expenditure (+/−)	Personal and public consumption expenditure (+)	Value of non-paid household labour (+)	Cost of non-renewable resource depletion (−)
		Value of volunteer labour (+)	Cost of agricultural land degradation (−)
	Defensive and rehabilitative expenditure (−)	* HK: Cost of crime (−)	Cost of fisheries depletion (−)
	Expenditure on consumer durables (−)	* S: Cost of security and external relations (−)	Cost of air pollution (−)
	Services from consumer durables (+)	Cost of unemployment and underemployment (−)	* HK: Cost of water pollution (−)
	Income distribution index (+/−)	Cost of overwork/lost leisure time (−)	* HK: Cost of noise pollution (−)
Services yielded from fixed capital (+)		Cost of family breakdown (−)	* HK: Cost of solid waste (−)
Change in net foreign assets (+/−)		* HK: Direct disamenity of waste-water pollution (−)	* S: Cost of environmental degradation (−)
			Cost of climate change (−)
		* HK: Direct disamenity of air pollution (−)	* HK: Value of carbon sequestration (+)
			Cost of lost wetland (−)

Note: Items marked '* HK' are only included for Hong Kong, items marked '* S' are only included for Singapore. All other items are included for both Singapore and Hong Kong.

Economic items

Weighted adjusted consumption expenditure (+/−)

The GPI begins with *Personal and public consumption expenditure*. However, that in itself does not reflect adequately the exact expenditure on welfare-yielding consumption. To estimate this, from *Personal consumption expenditure* we subtract the portion of *Expenditure* that is used for *defensive or rehabilitative* purposes, as well as *Expenditures on consumer durable*, while we add *Services from consumer durables*. This is then adjusted using an *Income distribution index* to obtain *Weighted adjusted consumption expenditure*, which consists in the bulk of the economic indicators.

Personal and public consumption expenditure (+/−)

The largest item in the GPI is usually personal and public consumption expenditure. In this study, we calculate *Personal and public consumption expenditure* as all private sector and public sector consumption expenditure. Since tobacco products and excessive alcohol consumption are not considered to improve well-being, in line with other studies we deduct all spending on cigarettes and tobacco and 25 per cent of spending on alcoholic beverages. Exactly what percentage of expenditure on alcoholic beverages (if any) should be deducted is of course open to debate. Clarke and Shaw (2008) deduct 50 per cent of the cost of alcoholic beverages, but in the Hong Kong and Singapore we feel that 50 per cent would be excessive.

Defensive and rehabilitative expenditures (−)

As part of the calculation of sustainable welfare, the differentiation between consumption expenditure that improves well-being, and other types of spending, is heavily emphasized. *Defensive and rehabilitative expenditures* are subtracted from *Personal and public consumption expenditure*, because they do not increase welfare, or well-being. There have been debates regarding the issue of what exactly should constitute *Defensive and rehabilitative expenditure*, because, as suggested by Neumayer (1999), ultimately all types of consumption would be counted as defensive or rehabilitative. For example, if health expenditures are considered defensive expenditures against illnesses, then expenditures on food and drinks could be included as defensive expenditures against hunger and thirst (Neumayer, 1999; Lawn, 2005). Although there is validity to these arguments, there is a clear difference between necessary expenditures, such as on food and drinks, and expenditures that people feel are necessary to protect themselves from the side effects of the economic processes, such as air purifiers from poor air quality, which could clearly be considered defensive and rehabilitative (Lawn, 2005). Additionally, expenditures spent on luxury products, such as on gourmet food, demonstrate that not all personal consumption expenditures should be considered defensive (Lawn, 2005).

To incorporate this item, studies by Lawn (2008a, 2008b) deduct a certain percentage of expenditures considered partly defensive or rehabilitative, such as food, health, education, insurance, household pollution abatement, vehicle accidents, and rent and dwelling services. Forgie *et al.* (2008) only accounts for defensive expenditure on health and education. On the other hand, Makino (2008) and Wen *et al.* (2008) do not specify how they calculate this item, while Clarke and Shaw (2008) do not include it.

Our study of Hong Kong incorporates the sub-categorizations found under private expenditure in the Hong Kong Annual Digest of Statistics. This study incorporates the method used in Lawn (2008a) and to some extent in Forgie *et al.* (2008). The former study accounts for a certain percentage of each sub-category of private expenditure as defensive or rehabilitative expenditure, while the latter study only accounts for defensive expenditures on health and education. The variations between our study and Lawn's (2008a) are only in the names of certain sub-categories. For example, the sub-category 'Spending on fuel and light' for Hong Kong corresponds to the sub-category 'Spending on electricity, gas and other fuel' used in Lawn (2008a). Therefore, the percentages we use to calculate the portion of expenditure deemed to be defensive or rehabilitative is also similar to Lawn's (2008a). For Hong Kong, *Defensive and rehabilitative expenditures* includes the following:

- 0.25 × spending on food (defensive);
- 0.25 × spending on rent and water charges (defensive);
- 0.25 × spending on fuel and light (defensive);
- 0.25 × spending on personal care (defensive);
- 0.5 × spending on health (defensive and/or rehabilitative);
- 0.25 × spending on transport and communications (defensive);
- 0.25 × spending on education and research (defensive);
- 0.125 × spending on recreation and entertainment (defensive);
- 0.5 × spending on financial and other services (defensive); and
- 0.25 × spending by general government (defensive and/or rehabilitative).

The items subtracted are slightly different for Singapore. We have included relevant items found under private expenditure in Singapore's Yearbooks of Statistics. However, due to nomenclature differences, some of the items have been renamed, for example 'rent and water charges' has become 'housing and utilities'. To estimate the GPI of Singapore, the following items were included:

- 0.25 × spending on food and non-alcoholic beverages (defensive);
- 0.25 × spending on housing and utilities (defensive);
- 0.25 × spending on personal care (defensive);
- 0.5 × spending on health (defensive and/or rehabilitative);
- 0.25 × spending on transport and communications (defensive);
- 0.25 × spending on education (defensive);
- 0.125 × spending on recreation and culture (defensive);
- 0.125 × spending on accommodation services (defensive);
- 0.125 × spending on food serving services (defensive); and
- 0.125 × spending on miscellaneous goods and services (defensive).

Expenditure on consumer durables (−)

Expenditures on consumer durables (such as furniture, furnishings, and household equipment) are deducted from personal consumption expenditure because the benefits arise over a period of time (these benefits are added under the item *Services from consumer durables*). *Expenditure on consumer durables* equals to the sum of all household expenditure on consumer durables.

Services from consumer durables (+)

The GDP adds the price of consumer durables in the year when the products are purchased. However, consumer durables are used for a number of years. *Services from consumer durables* reflects the value of the services yielded by the consumer durables during the lifespan of these products. Lawn (2008a) and Makino (2008) assume that consumer durables last ten years with a 7 per cent discount rate. Clarke and Shaw (2008) use the same number of years, but no discount rate. Wen *et al.* (2008) assume they last for ten years with a 10 per cent discount rate. Here we assume that stock endures for seven years. Hence, *Service from consumer durables* equal the value of consumer durables × 0.143 for each of seven years following the purchase.

Income distribution index (+/−)

Since income inequality is very high in many countries, including Hong Kong and Singapore, an income distribution index is used to gain a better understanding of people's economic status. The first four items of the *Economic Sub-index* (Table 4.1) are therefore aggregated and weighted with the *Income distribution index* to obtain the *Weighted adjusted consumption expenditure*. The *Distribution Index* is based on the change in income distribution over the study period as measured by the Gini coefficient (during the first year of the study period the index has a value of 100).

Services yielded from fixed capital (+)

This item includes services from non-dwelling infrastructure and capital goods provided by the private sector, publicly owned corporations, and the government. Publicly provided man-made capital includes infrastructural capital, such as roads, highways, bridges, schools, hospitals, and museums. Some studies, such as those of Wen *et al.* (2007) and Berik and Gaddis (2011), only include services from streets and highways. Studies that estimate the services yielded by all publicly provided man-made capital assume that 75 per cent of all government investment spending is devoted to fixed capital rather than producer goods; so the final value used to estimate the GPI consists of public-sector consumption of fixed capital multiplied by 0.75 (Lawn, 2008a, 2008b).

Private capital goods is the stock of built capital in the form of buildings and construction, factories, machinery, equipment, computer software, and tools that are used to create goods and services and that are not themselves used up in the production process (Fourie, 2006). This has to be maintained to sustain the level of production and consumption. This item has been estimated differently by different researchers. Makino (2008) and Clarke and Shaw (2008) assume that the services yielded from publicly provided man-made capital last for ten years with a 7 per cent discount rate. Wen *et al.* (2008) estimate this item as a percentage of Gross National Income multiplied by the ratio of government investment in public infrastructure to total investment in fixed assets. Forgie *et al.* (2008) estimated this item from government-owned capital stocks, calculated using the depreciation of capital stocks plus the opportunity cost of government investment funds. Berik and Gaddis (2011) included only services from streets and highways.

Many studies call this item 'Services yielded from publicly provided man-made capital'. Here, we rename the item *Services yielded from fixed capital* to better reflect what the data pertain to. In the 'Hong Kong Annual Digest of Statistics', data related to expenditure for infrastructure were only available in the category 'Gross domestic fixed capital formation'. The Yearbooks described gross domestic fixed capital formation as the 'gross value of investment expenditure on machinery, equipment and computer software, as well as building and construction; and costs of ownership transfer'. Similarly, the Singapore's Yearbooks of Statistics include the category of 'Gross fixed capital formation'. In order to reflect more accurately what the data pertain to, we rename the item 'Services yielded from fixed capital'. In this study we assume that the man-made fixed capital (including infrastructure) can be used for ten years, and multiply 'Gross domestic fixed capital formation' in the case of Hong Kong, and 'Gross fixed capital formation' in the case of Singapore, by 0.1, and include it for the ten years which follow the expenditure.

Net foreign lending or borrowing (+/−)

This item is included because 'a nation's long-term capacity to sustain the psychic income generated by the economic process depends very much on whether natural and human-made capital is domestically or foreign owned' (Lawn, 2003: 114). This can be either positive or negative. Studies have indicated that countries with large foreign debts would usually have trouble maintaining their natural and man-made capital intact, and are often forced to liquidate their natural capital stocks to repay debt (Lawn, 2003, 2005; Lawn and Clarke, 2008c). This item calculates the changes in foreign debt from year to year by allocating a base value of 100 to the first year of the period under consideration, and adjusting it yearly. Clarke and Shaw (2008) use changes in net foreign liabilities, while Makino (2008) uses 'net foreign lending', to better reflect Japan's situation. Lawn (2008a, 2008b), Forgie *et al.* (2008), and Wen *et al.* (2008) use changes in net foreign debt.

The consideration of foreign debt in other studies focused on understanding the implications of the added pressure of debt on the rate of liquidity of capital and assets. The same implications have been ascribed to net foreign assets, as these have also been related to a country's indebtedness (Wen *et al.*, 2008). As data covering Hong Kong's and Singapore's net foreign assets are available for a longer period of time than the figures for net foreign debt, we use changes in net foreign assets, and average changes over the previous five-year period.

Social indices

Value of non-paid household labour (+)

The contribution of non-paid household labour is not captured in the GDP. However, much credit should be given to their welfare-adding nature. Household labour consists in, among others, looking after and caring for family members, parenting and eldercare, as well as housework, such as the repairing and cleaning required to maintain the physical housing stock (Wen *et al.*, 2007; Lawn and Clarke, 2008c; Berik and Gaddis, 2011).

There are different methods to give an economic value of this item. The two most popular methods are the 'market replacement cost' methods and the 'opportunity cost method'. The 'market replacement cost' is based on the wage of a housekeeper. Forgie *et al.* (2008) calculate the cost using the median wage rate for housekeepers but adjust this through time for known changes in age–sex cohort demographics. Wen *et al.* (2008) also use the 'average hourly wage for resident services', but include technological improvements embodied in household appliances, which are assumed to reduce the number of household labour hours at the rate of 0.5 per cent per annum (Lu and Peng, 1999; Wen, 2005). The 'opportunity cost' method uses the income that could otherwise be earned if the person engaged in non-paid household labour worked for a wage (Jackson, and Styme, 1996; Brown *et al.*, 2011; Berik and Gaddis, 2011; Giannelli *et al.*, 2012). Here, we estimate the value of non-paid household labour using the net opportunity cost method.

We use data on the population that is aged 15 or above, is not participating in the labour force, not in an institution (e.g. prison), and not a student. A changeable percentage of males and females in this population are engaged in non-paid household labour, and we use the data of Wong (1996) and Ho (2007) to estimate that number. Hence, *Value of non-paid household labour* equals to the number of non-paid household labour hours, as estimated by Wong (1996) and Ho (2007) × average monthly wage in Hong Kong and Singapore.

Value of volunteer labour (+)

Volunteer labour provides services performed free-of-charge that would otherwise be performed by wage-workers, while helping strengthening social ties in a community (Berik and Gaddis, 2011; Sajardo and Serra, 2011). Volunteering provides psychological benefits to an individual, especially for those who are economically inactive (e.g. students and retirees) (Mjelde-Mossey *et al.*, 2002). Mjelde-Mossey *et al.* (2002) suggest that the psychological benefits of volunteerism include providing a way into active engagement in society, and describe the psychological benefits as a form of active aid to those in need and to those meeting the needs of others. Chou and Chi (2004) also point out that in an ageing society, social services become strained and volunteerism is one way to lessen this burden.

The number of volunteer workers is difficult to estimate. Wen *et al.* (2008) assumed that 70 per cent of all employees in China engage in volunteer work twice a year, on Arbor Day and Lei Feng Learning Day, as determined by law. Clarke and Shaw (2008) did not include this item, while Makino (2008) did not explain the method used. Here we estimate the value of volunteer labour using the net opportunity cost method. Statistics of volunteer hours are available from the Central Office of Volunteer Service, Social and Welfare Department, Hong Kong, and from the Singapore's Yearbook of Statistics. Here, *Value of volunteer labour* equals the number of volunteer labour hours × assumed wage. In Hong Kong we estimate the value of volunteer labour using the minimum wage rate of $28 in 2010 (and discount it for previous years). In Singapore the wage rate used corresponds to the average monthly wage in the community, social and personal services industry.

Cost of crime (−) (only for Hong Kong)

Crime negatively affects welfare. The cost of crime includes the value of the property stolen or damaged, the cost of medical expenses, the value of lost wages, the cost of policing, prosecuting and punishing the criminals, as well as the social costs that crime has towards human relations, social institutions, and the self-esteem of citizens (McCollister *et al.*, 2010). However, both Lawn and Clarke (2008c) and Berik and Gaddis (2011) have noted that studies should be cautious about the type of crimes included in the calculation, as crime costs may already be reflected in other GPI indices. For example, the costs associated with arson and sabotage would be reflected in lower consumption expenditure. The cost of crime would also be reflected through higher values of defensive expenditures (such as on fire prevention or burglary alarms) (Lawn and Clarke, 2008c). For this reason, defensive expenditures regarding crime prevention are not included into the cost of crime. Wen *et al.* (2008) based their estimates on insurance premiums paid to property insurance companies and expenditure for both the legal system and law enforcement. On the other hand, Lawn (2008a, 2008b), Forgie *et al.* (2008), Makino (2008), and Berik and Gaddis (2011) estimated it by multiplying the crime rate by the cost for each crime category. Clarke and Shaw (2008) do not include this item.

For Hong Kong, *Cost of crime* is calculated according to total expenditure on the insurance premium paid for property damage, pecuniary loss, 80 per cent of expenditure on law enforcement (80 per cent as defensive expenditure and 20 per cent as necessary for law and order of the city), less the total value of property stolen and recovered. Here we exclude medical expenses and lost wages, and other defensive expenditures. Because of lack of data, this item is not included for Singapore. Instead, for Singapore we use *Cost of security and external relations*.

Cost of security and external relations (−) (only for Singapore)

This item is only used for Singapore, because *Cost of crime* is not reported in the Singapore's Yearbook of Statistics. *Cost of security and external relations* is calculated using total expenditure for justice, order and security. The cost is calculated in relation to the total expenditure of Singapore's Ministry of Defense, which combines operating and development expenditure.

Cost of unemployment and underemployment (−)

Unlike the GDP, the GPI includes the *Cost of unemployment and underemployment*. Mainstream macroeconomics states that a certain rate of unemployment is necessary to keep under control the rate of inflation. The 'non-accelerating inflation rate of unemployment' (NAIRU) involves 'reducing aggregate demand through monetary policy settings (often higher interest rates) in order to allow unemployment to rise sufficiently to achieve an inflation-controlling ratio of unemployed labour to conventional workers' (Lawn and Clarke, 2008c; Lawn, 2010). Lawn (2010) has described this as 'an insidious means of controlling inflation since it requires the permanent existence of a sacrificial pool of unemployed labour'.

Unemployment and underemployment are both signs of the inefficiency of the economic system to utilize the human capital stock a nation has, and reflect an undesirable side effect of the market economy. It is important to also include underemployment as a social cost. Those who are chronically unemployed or have given up looking for jobs (i.e. discouraged workers) or those who work fewer hours than they would like to (e.g. part-time workers who would like to work full time) should also be included among the unemployed and underemployed (Berik and Gaddis, 2011).

The costs involved include the loss of potential output (in terms of income and capital), reduced consumption and personal freedom, social exclusion, negative impacts on relationships and family life, deterioration of skills and motivation to work, degradation of one's psychological state, ill-health and reduced life expectancy, increased racial and gender discrimination in respect to income and wealth differentials, and increased rates of suicide (Lawn and Clarke, 2008c; Berik and Gaddis, 2011). However, it must be noted that some of these costs could be reflected in other indices, such as in reduced consumption expenditure, increased costs of crime or family breakdown, and increased defensive or rehabilitative expenditure (Lawn and Clarke, 2008c). Therefore, the general calculation depends largely on the estimated number of 'unprovided' hours due to unemployment and underemployment, where unprovided hours of underemployment is the difference between the average hours of underemployed workers and the average hours worked by full-time workers (Lawn, 2008a/b?; Forgie *et al.*, 2008; Wen *et al.*, 2008; Makino, 2008; Lawn and Clarke, 2008c; Berik and Gaddis, 2011). Here, we estimate *Cost of unemployment and underemployment* by multiplying the population of unemployed and underemployed, as reported by the statistical yearbooks × average annual income.

Cost of overwork and lost leisure time (−)

Overwork and lost leisure time have adverse effects on welfare (Golden and Wiens-Tuers, 2008). These activities are likely to result in additional output, which means that they increase GDP. However, in the GPI they are deducted because overwork results in reduced welfare. There is an interesting dilemma here when attempting to estimate how much leisure time should be regarded as a benefit to welfare. Firstly, there is no doubt that fully employed workers who experience overwork would treat leisure time as a clear benefit. However, underemployed workers would point out that they work fewer hours than they desire and, therefore, are 'enjoying' certain hours of forced leisure (Wen *et al.*, 2007; Berik and Gaddis, 2011).

The cost of overwork or lost leisure time would ideally be calculated by subtracting the amount of unpaid and paid overwork hours from the average working hours the workers would prefer. Hence, estimating the value of this cost requires estimating the number of working hours (the base working hours) that are desired by both fully employed and underemployed workers. In Australia this has been estimated to be 37.5 hours per week, in the US state of Utah 35 hours per week, in Japan 35 hours, and in New Zealand 23.8 hours (Hamilton, 1999; Forgie *et al.*, 2008; Makino, 2008; Berik and Gaddis, 2011). With the base working hour defined, the amount of overwork hours can be estimated, and each additional (unpaid) work hour multiplied by the average hourly income (Lawn, 2008a; Forgie *et al.*, 2008; Makino, 2008; Berik and Gaddis, 2011). The total cost of overwork equates to the number of lost leisure hours of all fully employed workers multiplied by the leisure-adjusted real wage per hour (Berik and Gaddis, 2011). Lawn (2008a) adjusted the cost of lost leisure time upwards by a factor of 1.28, arguing that the value of leisure time is higher than the value of working time. Wen *et al.* (2008), and Clarke and Shaw (2008) do not include this item.

Hong Kong and Singapore have a long tradition of overwork. Overwork is a norm and a sign of commitment to the company, despite a general acknowledgement of its undesirability. Labour and trade unions in Hong Kong have proposed passing a law on maximum working hours, which would restrict the number of working hours an employee should work for. The maximum working hours proposed have been set to 44 hours per week, while research has indicated that on average employees work 10 more hours than this every week (HKU, 2008). Our data of Hong Kong are based on the total number of hours of overwork found by HKU's (2008) study, which states that about 62.5 per cent of employed people work for 10 extra hours per week. We calculate the cost adjusting the hourly wage upwards by 1.28, and multiplying this leisure-adjusted hourly wage rate by the extra working hour above a 'normal working hours' (estimated to be 44 hours a week) by fully employed workers (those that work at least 35 hours per week for 50 weeks per year). For Singapore, we do not have survey data as accurate as those for Hong Kong, and use a 2011 survey by Regus (Regus, 2011) to estimate the additional number of extra hours worked weekly. As in the case of Hong

Kong, the average hourly rate (multiplied by 1.28) is used to give a monetary value to the item.

Cost of family breakdown (−)

The family unit is a very important social institution that provides individuals with a secure, stable and organized environment, as well as one that serves crucial child-rearing functions (Lawn and Clarke, 2008c). Family breakdown is one of the consequences of stress and anxiety associated with contemporary social life, and is partly a side effect of an unhealthy economic system that fails to balance social life and environmental integrity with economic prosperity. Family breakdown is sometimes caused by trying to maintain or achieve particular consumption patterns by working harder, and perhaps taking on debt to sustain a particular standard of living (Berik and Gaddis, 2011). The results are family disunity and dysfunctionality, which often translate into increased numbers of divorce cases and the 'tag-along' costs associated with divorce, including legal fees, medical expenses, and the traumatic implications for the spouses and the children involved (Bougheas and Georgellis, 1999; Wen *et al.*, 2007; Lawn and Clarke, 2008c).

Calculating this index requires estimating the number of divorces and the number of children affected by divorces, per annum. Makino (2008) also included the cost of suicide. Wen *et al.* (2007) and Berik and Gaddis (2011) based the costs on the estimated sum of legal fees, counselling, and costs of relocating to separate households, money which could otherwise have been spent on welfare-adding activities. The cost per child affected by divorce varies by study. For example, the cost of divorce to children in China is estimated to be Yuan 2,200 per child (Wen *et al.*, 2007).

Here, the costs are based on changes in the number of divorce decrees with legal expenses and cost of lawyers, weighted according to the estimated number of children affected (based on average domestic household size). In other words, *Cost of family breakdown* equals to (total number of divorce decrees × cost of legal fees) + (total number of divorce decrees × cost of lawyer), weighted by number of children affected.

Direct disamenity of waste-water pollution (−) (only for Hong Kong)

Direct disamenity of waste-water pollution is an item other researchers have not included. We justify its inclusion by the fact that these costs are potentially high for city-states surrounded by the sea, as is the case for Hong Kong and Singapore.[1] The cost of waste-water pollution is based on the capital expenditure and other expenditures dedicated to water treatment, such as sewage treatment and harbour cleaning (assuming capital endures for ten years) and public expenditure on the environment (adjusted to inflation). We assume that *Direct disamenity of waste-water pollution* corresponds to 20 per cent of the total wastewater treatment costs, as reported in the Hong Kong Annual Digest of Statistics. The census data of

Singapore do not report this index, and we use the *Cost of environmental degradation* instead (see below).

Direct disamenity of air pollution (−) (only for Hong Kong)

This item reflects the reduced aesthetic aspect of, and immediate irritation caused by, air pollution, such as reduced visibility and difficulty breathing (as opposed to its long-term health effects accounted for in the *Cost of air pollution*). This cost has only been included in Lawn (2008a), where it accounts for 20 per cent of total air pollution cost. This index is particularly relevant to Hong Kong and Singapore, because of the high level of air pollution in these cities (see the relevant sections on air pollution in Chapter 5 for Hong Kong, and Chapter 6 for Singapore).

Direct disamenity of air pollution is estimated using capital expenditure on air-related projects (assuming capital endures for ten years) and public expenditure on the environment, as well as lost productivity from premature deaths caused by air pollution (reflected by the loss of five years of per capita GDP as a result of premature deaths). Hence, *Direct disamenity cost of air pollution* equals to 0.2 × (total air pollution cost + lost productivity from premature deaths due to air pollution). Unfortunately the census data of Singapore do not report this index, and as with the case of *Direct disamenity of waste-water pollution* we are forced to use the *Cost of environmental degradation* instead (see below).

Environmental items

Cost of non-renewable resource depletion (−)

Non-renewable resources included in the calculation range from fossil fuels (coal, natural gas, petroleum) (Wen *et al.*, 2007; Berik and Gaddis, 2011) to mineral resources such as copper, iron, and zinc (Wen *et al.*, 2007). According to both the Hicksian and Fisherian concepts of income and capital (Chapter 3), a portion of income generated from extracting non-renewable resources should be regarded as a cost, and therefore deducted from the national income, because non-renewable resources cannot be sustainably exploited (they cannot regenerate for future consumption) (Martinet, 2012). In addition, the costs of lost natural capital services – lost resources, 'sinks' (the ability to absorb waste, such as carbon dioxide), and life-support services provided by natural capital – should be deducted from national income to maintain the throughput of energy and materials required to keep the stock of man-made capital intact (Lawn, 2005; Lawn and Clarke, 2008c).

There have been numerous approaches to calculate this item. For example, Wen *et al.* (2007) estimated the cost of non-renewable resource depletion using the wholesale prices of energy and minerals. Cobb *et al.* (1995) only estimated the costs of depletion of non-renewable resources related to energy, excluding minerals, because of the 'importance of energy as an input and the ease of aggregating across energy resources'. However, most studies have used two approaches when estimating the costs of this item: the production-driven method and the consumption-driven method. The choice of method most suitable for calculating the cost has been debated and discussed in Neumayer (2000) and Lawn (2005).

The resource rent approach ('production-driven method')

The resource rent approach captures the loss of resources and associated wealth to future generation, based on the value of non-renewable resources produced in a year in a particular area. It is also referred to as the 'production-driven method' (Berik and Gaddis, 2011). Usually the resource rent approach involves the deduction of the total cost of non-renewable resource depletion. This is the method originally employed by Daly *et al.* (1989) and Cobb and Cobb (1994), who deducted the total value of mine extraction in their original ISEW computations. However, the revised ISEW in Cobb and Cobb (1994) 'switched to the replacement cost method: for non-renewable energy extraction each barrel of oil equivalent was valued at a replacement cost' (Neumayer, 2000: 349).

The replacement cost approach ('consumption-driven method')

The replacement cost approach values the permanent loss of wealth associated with the resources consumed, regardless of where they are produced. It is also referred to as the 'consumption-driven method' (Berik and Gaddis, 2011). Neumayer (2000) believed that the resource rent approach reflects the cost most accurately, arguing that there is no reason why non-renewable resources should be fully replaced when there are currently reserves available for future years (Lawn, 2005). Similarly, Lawn (2005) and Lawn and Clarke (2008c) argued that it is only necessary to deduct a portion of the value of non-renewable reserves. Neumayer (2000) stated that El Serafy's (1989) 'user cost' formula was the correct method of calculating resource rents, as the 'user cost' reflects the portion needed for investment and is deducted (Lawn, 2005).

Hamilton (1999) estimated Australia's cost of non-renewable resource depletion using the replacement cost approach, whereby he estimated the cost of replacing fossil fuels and nuclear energy with renewable alternatives, focusing particularly on the cost of producing 'gasohol' from biomass. Lawn (2005) believes that the replacement cost approach is more appropriate than the resource rent approach. In line with the concept of sustainability, the replacement cost approach is necessary to determine the amount needed (i.e. the portion that must be set aside to maintain the flow of income, namely the 'user cost') to fully replace non-renewable resources, so as to sustain future economic activities, and maintain future levels of production and consumption. The replacements considered in calculations are mainly renewable resources, whose costs can be fairly easily estimated (Lawn, 2005). Further, even though replacement is currently unnecessary, the actual cost of establishing substitutes to renewable resource may have to be attributed to the point in time when the depletion took place. Indeed, this is the basis behind El Serafy's user cost method (Lawn, 2006).

El Serafy's user cost method

El Serafy's (1989) user cost formula is as follows:

$$X / R = 1 - [1 / (1+r)^{n+1}]$$

Where X = true income (resource rent);
R = total net receipts (gross receipts less extraction costs);
r = discount rate;
n = number of periods over which the resource is to be liquidated;

Using this formula, $R - X$ corresponds to the user cost or the amount of total net receipts that must be set aside to establish a replacement asset to ensure a perpetual income stream. Therefore the user cost can be rearranged to become:

$$R - X = R[1 / (1+r)^{n+1}]$$

El Serafy's user cost method can be used to calculate both the production- and the consumption-driven methods, since X (resource rent) is 'the portion of proceeds from resource extraction that does not constitute a user cost', and $R - X$ (user cost or replacement cost) is 'the portion of the proceeds from resource extraction that does constitute a user cost [and] is, in fact, the genuine cost of resource asset replacement' (Lawn, 2005).

The number of periods over which the resource is to be liquidated depends on the size of a non-renewable resource deposit and the expected price of its resources (Lawn and Clarke, 2008c). The non-renewable resource price would also need to reflect the increasing absolute scarcity of the resource (absolute scarcity reflects the scarcity of the total quantity of all low-entropy resources available for current and future production, as opposed to relative scarcity, such as the scarcity of oil relative to coal). Many studies have estimated this rise in non-renewable resource prices with a 3 per cent escalation factor (Lawn, 2005). In terms of determining the most appropriate discount rate, Lawn (2005) suggested that the chosen discount rate should be equivalent to the real interest rate of the renewable resource, which might be its natural regeneration rate.

The *Cost of non-renewable resource depletion* should be calculated over the years as a cumulative total, which means that the cost should equal the cost incurred in a particular year as well as the sum of the costs accumulated from previous years. Because the GPI aims to evaluate the sustainable welfare enjoyed by a nation's citizen in a particular year, which is influenced by past activities, the cost of the depletion of non-renewable resources would eventually be borne on into the future (Lawn, 2005). Therefore, the total cost should reflect the amount that is required to compensate a nation's citizens in that particular year for the costs of current non-renewable resource depletion as well as the additional cumulative impact of past non-renewable resource depletion (Lawn, 2005).

However, one can also make the counter argument that only those natural assets that cannot be substituted by other man-made capitals or other resources – such as the global climate system and stratospheric ozone, should be accumulated (Lawn and Clarke, 2008d) – while natural resources that can be replaced – such as non-renewable energy (e.g. coal) replaced by renewable energy (e.g. solar energy) – do not require the accumulation of costs over time. Thus, the costs of timber and non-renewable resource depletion should not be accumulated. Instead, only the user cost incurred during the current year should be accounted for (Lawn and Clarke, 2008d).

Both arguments have strengths and weaknesses. The first argument suggests inter-generational equity should be taken into account, whereas the second upholds the concept of strong sustainability – the non-substitutability of certain natural capital. It can be argued that every cost bears with it intergenerational influences. For example, the cost of family breakup could mean a future cost of lower productivity for the children, and the impact of pollution could have negative consequences on the physical development of infants. In addition, keeping in mind that the GPI value of each year measures the well-being of a nation's citizens in that particular year, any cost incurred to future generations will be reflected directly or indirectly in future GPI values. We contend that the accumulation of costs should only be used for those natural capitals that are not substitutable (i.e. where no man-made capital, even by way of technological advances, and no other natural capital, can adequately replace the lost natural capital). In terms of the cost of non-renewable resource depletion, accumulation of cost should not be employed when there are renewable alternatives, as reflected by the user cost and replacement cost approaches.

Here, we use the replacement cost (using El Serafy's user cost formula) to estimate the amount that must be set aside to sustain a flow of income equal to that generated by the exhausted resource. In line with other studies, we assume that a mine (when the non-renewable resource is mined) is in operation for 31 years and the regeneration rate of the replacement assets is 1 per cent per annum. This implies that the set-aside component from non-renewable resource extraction constitutes 27.3 per cent of the net receipts obtained. On the assumption that market prices undervalue the absolute scarcity of non-renewable resources by half, we double the user cost percentage to 55 per cent. For Hong Kong, the cost is based on an average of US$100 for each tonne of local production of feldspar, kaolin, quartz and other quarry minerals, according to Hong Kong Census and Statistics. For Singapore, the cost is based on an average of US$100 for each tonne of local production of metallic ore (from the Department of Statistics Singapore). For both Hong Kong and Singapore the *Cost of non-renewable resource depletion* equals to the total value of non-renewable resource output × 0.55.

Cost of agricultural land degradation (−)

The *Cost of agricultural land degradation* (called *Cost of lost farmland* in some studies) aims to reflect the amount of money needed to compensate for the loss of fertile agricultural land as a result of unsustainable agricultural practices. In Berik and Gaddis (2011) this index is renamed 'option value of farmland', with the argument that instead of being treated as a cost, the option value is the value of preserving the option to use the ecological services that the land can provide, in this case the service of food production. In our opinion, the loss of farmland should be best reflected as a cost because its additional value may well include the psychic income of certain services, especially in crowded cities like Hong Kong and Singapore. Also, calling this cost an option value would easily overstate overall GPI, as option values are psychic incomes not enjoyed physically.

There are numerous methods of calculating this cost. In the study of Australia's GPI, the calculation reflects the amount required to compensate citizens for the cumulative impact of past and present agricultural practices, which is equal to the estimated annual cost of land degradation weighted by the area of land used for agricultural purposes annually (Lawn, 2008a). This cost is accumulated to satisfy the concept of strong sustainability – agricultural land as irreplaceable natural capital (Lawn, 2008a). In the studies of India's and Thailand's GPI, the annual cost is equal to 1 per cent of the value of agricultural outputs, and is accumulated over time (as it is in the study of Australia) (Clarke and Shaw, 2008; Lawn, 2008b). The same method has been used for the study of Vietnam's GPI, but instead of 1 per cent, Nguyet Hong *et al.* (2008) used 7 per cent of the value of agricultural output to estimate the annual cost of land degradation. In the study of Japan's GPI, the cost of lost farmland has been calculated using the costs of the lost ecological and social functions they provide (prevention of floods, land erosion and mudslides, absorption of organic wastes, air purification, climate stabilization, recreation and relaxation, and a source of food production) (Makino, 2008). Makino regards the multi-functionality of farmland as non-paid benefits generated by its sink and ecological functions, and estimates the annual costs of lost sink and ecological services. In the study of China's GPI, the cost is treated as the amount needed to compensate for the accumulated area of lost agricultural land, which includes the decline in agricultural productivity, the opportunity cost of rehabilitative expenditures incurred to restore damaged farmland, and the loss of ecological services (Wen *et al.*, 2008). The total cost is calculated by multiplying the estimated cost of each unit of lost agricultural land to the cumulative loss of agriculture land. Some studies, such as that of New Zealand, exclude this index (Forgie *et al.*, 2008).

The cost of agricultural land degradation may be accumulated over time because land degradation is non-substitutable (Lawn, 2008a, 2008b; Clarke and Shaw, 2008). However, considering that degraded land could eventually recover its former productive capacity if the land is left to fallow or restored, accumulating the costs indefinitely would be misleading. Therefore, Lawn (2008a) adopted the approach of weighting the cost downwards with a recovery factor of 1 per cent per annum.

The importance of agriculture in Hong Kong and Singapore has greatly diminished over the past four decades because the demand for food crops has increasingly been met by imports, the pool of farmers has diminished, and farmland has been used for urban development. To estimate the cost of agricultural land degradation we use the approach of Nguyet Hong *et al.* (2008), and assume that 7 per cent of total agricultural output needs to be set aside for restoration purposes. Also, we do not aggregate the costs of previous years because agricultural practices in Hong Kong and Singapore are low-intensive. Thus, *Cost of agricultural land degradation* equals to 0.07 × total value of agricultural outputs (local production of vegetables, fruit, rice, and field crops multiplied by their respective current retail value).

Cost of fisheries depletion (−)

In the past, Hong Kong and Singapore hosted important fishing industries in their territorial waters. Now, fishing has considerably shrunk because of overfishing, and alternative, more attractive, employment opportunities. The cost of fishery depletion is estimated using the replacement cost from El Serafy's user cost formula, as the amount needed to set aside to sustain a flow of income equal to that generated by the exhausted resource. In line with other studies, we assume that captured fisheries last 24 years and the regeneration rate of the replacement assets is 1 per cent per annum. This implies that the set-aside component from fishery resource extraction constitutes 22 per cent of the net receipts obtained. In line with other studies, we also assume that the market prices undervalue the absolute scarcity of fishery resources by half, and double the user cost percentage to 44 per cent. Hence, *Cost of fisheries depletion* equal to production value of captured fisheries (salt and fresh-water fish) × 0.44.

Cost of air pollution (−)

The high (and rising) cost of air pollution originates partly from the growing amounts of pollutants emitted, and partly from the loss of the natural environment's waste-assimilative or sink capacity. The largest costs of air pollution are related to the impact on human health (premature mortality and increased frequency of illnesses) caused by harmful substances and fine airborne particulate matters, in terms of decrease in lung function, increase in heart attacks, and worsening of asthma (Fann and Wesson, 2009; Berik and Gaddis, 2011). Not only does this harm human health but it also means that larger portions of resources have to be set aside for the maintenance of human capital. In addition to health costs, other costs associated with air pollution include a reduction of recreational values due to poorer visibility, and an impaired ecosystem (Hyslop, 2009). These are considered in another item, *Direct disamenity of air pollution* (a Social index discussed in the previous section).

Different studies have used different approaches to estimate this cost. Berik and Gaddis (2011) calculated it by obtaining information on the six largest pollutants (large particulate matter (PM10), fine particulate matter (PM2.5), nitrogen oxides (NOx), sulphur oxides (SOx), and volatile organic compounds (VOCs)), and estimated the costs on health and visibility, as well as the loss of economic development opportunities that are associated with companies relocating to other cities due to poor air quality. Lawn (2008a) used 'loss of sink capacity', and estimated it to correspond to 80 per cent of the total air pollution cost based of sulphur oxide (SOx) and nitrogen oxide (NOx) emissions. Lawn (2008b) estimated the costs of air pollution using 'real output factor' (assumed to be 2.5 per cent of real GDP) multiplied by a pollution abatement technology factor with an assumed annual rate of progress of 2 per cent. Clarke and Shaw (2008) estimated the cost of air pollution by equating it to the total expenditure of pollution abatement from carbon dioxide, carbon monoxide, NOx, SOx, and suspended particulate matter. Makino (2008) estimated the costs by multiplying the levels of sulphur oxide and nitrogen oxide emissions with their relevant marginal costs. Perhaps because of the absence of detailed information on particular pollutants, Wen *et al.* (2008b) used the World Bank 2007 estimates that the costs of air pollution account for 3.8 per cent of GDP, and used 50 per cent of that amount assuming that the remaining 50 per cent were already reflected in other GPI items, such as the loss of working days and costlier defensive or rehabilitative expenditure. In the same vein, Nguyet Hong *et al.* (2008) estimated the cost of air pollution by multiplying the annual expenditure on air pollution mitigation by five, assuming that the total cost of air pollution is five times the mitigation cost. The *Cost of air pollution* does not need to be accumulated over time, because air is 'unlimited'. However, to ensure intergenerational equity, lowering the cost of air pollution would require reduction of emissions and stringent emission policies.

Air, waste-water, noise, and solid waste pollution all fall within the scope of the Hong Kong government's environmental policy, as the four areas are specifically categorized in the Hong Kong Annual Digest of Statistics under 'Capital expenditure devoted to environmental projects by type'. The *Cost of air pollution* is calculated based on lost productivity due to premature death by cardio-respiratory diseases caused by air pollution. This is multiplied by the per capita GDP of each respective year, which is then multiplied by five on the assumption that premature death reduces a person's life on average by five years. Quah and Boon's (2003) used a similar approach to estimate the cost of premature death for Singapore. However, Quah and Boon (2003) used the concept of Value of Statistical Life (VOSL), whereas in our study we refer to GDP per capita because we believe that the GDP per capita better reflects the opportunity costs associated with lost productivity or leisure time. Furthermore, the VOSL assumes that the victim would have a lifespan's worth of productivity, and consequently overvalues such costs. Only $PM_{10}{}^2$ is included in our study, because studies have suggested that it is the most harmful of all air-pollutant particles, and is the main cause of lung cancer and premature deaths (Raaschou-Nielsen *et al.*, 2013). The number of premature deaths from cardio-respiratory diseases given in the Hong Kong Annual Digest of Statistics is multiplied by 0.01 on the assumption that

1 per cent of the deaths were caused by the pollutant PM_{10}. We leave out the costs of morbidity to avoid double-counting the defensive and rehabilitative component of the study. However, we add the capital expenditure on air pollution, which includes such things as installing additional air pollution monitors. Each capital expenditure is assumed to last for ten years (as was the case with services from infrastructure). We multiply the final figure by 0.8, to retain Lawn's (2008a) concept of 'loss of sink capacity', which he estimated to correspond to 80 per cent of the total air pollution cost.

For Singapore, only the costs of premature mortality due to air pollution are included because of lack of data on government expenditures to address the problem of air pollution. The expenditures that the Singapore government undertakes to address air pollution, waste-water pollution and noise pollution are included under the *Cost of environmental degradation* item, below. We follow the methodology used by Quah and Boon's (2003) study, with the different assumption that premature deaths reduces on average five years' worth of productivity. For Singapore, the *Cost of premature mortality due to air pollution* equal to 5 × GDP per capita × estimated number of premature deaths due to PM10.

Cost of water pollution (−) (only for Hong Kong)

Similar to the *Cost of air pollution*, the *Cost of water pollution* reflects the loss in the natural environment's waste-assimilative or sink capacity due to the increasing quantity and worsening quality of waste generated by economic activities (Lawn and Clarke, 2008c). Clean water in lakes, streams and oceans provides ecological services such as clean drinking water, healthy fisheries and aquatic life, pristine recreational areas and aesthetic qualities, and increases property values (Postel and Richter, 2003). The cost of water pollution consists in the loss of the natural environment's waste-assimilative or sink capacity, increasing costs for water treatment, and losses to the tourism industry and property values (Gibbons, 2013).

A variety of methods have been used to calculate this cost. Berik and Gaddis (2011) based their estimate on several quality impairments associated with water pollution: to drinking water, to recreational uses, and to aquatic life. Lawn (2008a) estimated the cost of waste-water pollution using information on the cost of waste-water treatment for urban Australians, multiplied by the number of urban Australians. Makino (2008) based his estimates on the levels of biological oxygen demand (BOD) and chemical oxygen demand (COD), disregarding pollutants such as phosphate, nitrogen, and heavy metals. Forgie *et al.* (2008) used the cost of riparian planting of lowland river margins, and planned restoration work on eutrophic lakes. Lawn (2008b) estimated the annual cost to be equivalent to the real output factor (corresponding to 2.5 per cent of real GDP) multiplied by a 'pollution abatement factor' (assuming a 2 per cent annual rate of progress) and the water withdrawal factor. Nguyet Hong *et al.* (2008) assumed that the cost of waste-water pollution corresponds to five times the annual cost of mitigation of urban waste-water pollution. Clarke and Shaw (2008) assumed that the cost equals the cost of rehabilitating polluted urban waste-water, and multiplied it by two to account for non-point sources of waste-water pollution. Lastly, Wen *et al.* (2008) assumed the cost to be 50 per cent of the total cost of water pollution, where the total cost was the sum of the investment to implement waste-water treatment and the cost of the impact on human health.

Like for *Cost of air pollution*, and unlike *Cost of agricultural land degradation*, the *Cost of water pollution* does not require the yearly accumulation of costs because water is abundant. However, lowering the cost of water pollution would require more stringent water monitoring and control policies. For Hong Kong, we have estimated the *Costs of water pollution* using the total 'cost of waste-water pollution' reported in the Hong Kong Annual Digest of Statistics under 'Capital expenditure devoted to environmental projects by type' as the benchmark. For Hong Kong, *Cost of water pollution* equals to 0.8 × total waste-water cost. For Singapore, this item is not included because we are unable to find sufficient information. Instead, we use *Cost of environmental degradation.*

Cost of noise pollution (−) (only for Hong Kong)

The Cost of noise pollution reflects the immediate irritation caused by noise pollution, based on the public expenditure devoted to noise-related environmental projects. The *Cost of noise pollution* equals to expenditure on noise-related environmental projects. We use the data from the Hong Kong Annual Digest of Statistics under 'Capital expenditure devoted to environmental projects by type', and assume capital endures for ten years. Hence, we estimate *Cost of noise pollution* as 0.1 × total capital expenditure to address the cost of noise pollution, for each of the ten years that follow the year the investment was made.

Cost of solid waste (−) (only for Hong Kong)

The cost of solid waste could include the net opportunity cost of solid waste that ended up in landfills or incinerators but which could have gone to recycling, and the lost source of income to grass-root communities that could have been engaged in that recycling. Because of lack of information on these potential economic opportunities, we estimate the *Cost of solid waste* based on capital expenditure related to solid waste, as reported in the Hong Kong Annual Digest of Statistics under 'Capital expenditure devoted to environmental projects by type'. We assume that capital endures for ten years. Hence, we estimate *Cost of solid waste* as 0.1 × total capital expenditure to address the cost of solid waste, for each of the ten years that follow the year the investment was made.

Cost of environmental degradation (−) (only for Singapore)

In contrast to Hong Kong, where government expenditure to address different environmental problems is broken up, Singapore's census data only reports one total amount for all problems under *Cost of environmental degradation*. This exposes one limit of the GPI: the lack of sufficiently specific data for some countries. The *Cost of environmental degradation* is calculated by computing the total expenditure, which consists of operating expenditure and development expenditure, of the environmental projects implemented by Singapore's Ministry of Environment and Water Resources. These projects aim to create a cleaner, more sustainable environment for the population, and promote environmental awareness. Operating expenditure refers to the expenditure to operate or run the projects, such as manpower and land costs. Development expenditure refers to capital expenditure on ongoing projects, and construction sites that are under development and are assumed, for the purpose of this calculation, to last for ten years. We estimate *Cost of environmental degradation* as $0.1 \times$ total operating expenditure and development expenditure, for each of the ten years that follow the year the investment was made.

Cost of climate change/Cost of long-term environmental damage (−)

Climate change results from the emission of large amounts of greenhouse gases, a side effect of economic activities. Some of the side effects of climate change are loss of the world's biodiversity, increased frequency of drought, and increased frequency and intensity of extreme weather events, such as typhoons (Lawn and Clarke, 2008c; Berik and Gaddis, 2011). This cost is related to the level of carbon dioxide, as this is the main anthropogenic greenhouse gas. Some studies call this item differently, while others combine it with the cost of long-term environmental damage (for example, Forgie *et al.* (2008), Makino (2008), Wen *et al.* (2008)).

The calculation of this item mainly focuses on the amount of carbon dioxide-equivalent emissions associated with energy production and/or the production of goods and services. The exact approach, however, varies across different studies. Berik and Gaddis (2011) calculate the cost using the amount of carbon dioxide-equivalent emissions associated with personal consumption in metric tons per year. Other studies use similar methods, calculating the costs borne by society to produce and consume a bundle of goods and services. Wen *et al.* (2008) take into consideration CO_2 emissions (multiplied by an unspecified percentage of Gross National Income) and ozone-depleting substances (giving a value of US\$15.26 per kilogram produced), and divide the value found by half, arguing that some of the long-term costs of climate change are already reflected in other GPI items. Forgie *et al.* (2008) estimate the cost of climate change by multiplying the level of greenhouse gas emissions by the price of carbon of the European Union Emissions Trading System (EU-ETS). Clarke and Shaw (2008) give a price of Baht 21.59 per tonne of carbon emissions, accumulating the cost but weighting it downwards by a 1 per cent recovery factor. Lawn (2008a, 2008b) considers the amount required to compensate future citizens for the long-term environmental impact of energy consumption. Makino (2008) calculates the cost of climate change by multiplying the cumulative carbon dioxide emissions by the marginal social cost of emissions each year – the marginal social cost being the cost society has to bear for economic activities.

The costs of climate change accumulate over time since climate cannot be substituted by man-made capital. However, if global carbon dioxide levels are stabilized, the world's climate system is also expected to stabilize (Lawn and Clarke, 2008d). Therefore, it can be argued that the cost of climate change should not be added indefinitely. Lawn (2008a) tackles this problem by weighting the cost downwards by a recovery factor of 2 per cent per annum, while Clarke and Shaw (2008) adopted a 1 per cent per annum recovery factor.

Here, we estimate the *Cost of climate change* based on the amount and global warming potential (GWP) of six different greenhouse gases (carbon dioxide (CO_2), methane (CH_4), nitrous dioxide (N_2O), hydrofluorocarbons (HFCs), perfluorocarbons (PFCs), and sulfur hexfluoride (SF_6)). The global warming potential is based on the Fourth Assessment Report of the Intergovernmental Panel on Climate Change (IPCC). The price of carbon is based on carbon market value (US$20 per tonne of carbon dioxide equivalent in 2009). We expect a recovery rate of 2 per cent per annum.

Hence, the *Cost of climate change* equal to amount of emissions of each GHG (tonnes) × its global warming potential (GWP) × cost per carbon dioxide equivalent tonne × 0.98 recovery rate + cumulative cost from previous years.

Value of carbon sequestration (+) (only for Hong Kong)

Growing awareness of the harm caused by raising amounts of greenhouse gases in the atmosphere has created a market for carbon, and spurred investments in carbon sequestration (Van Kooten *et al.*, 2012). Most studies do not include this item, but we feel that if we include the cost of climate change, it is only fair to also include the value of the carbon sequestered. The greatest sinks of atmospheric carbon are water bodies (principally the oceans), soils and plants. Here, the *Value of carbon sequestration* is estimated using the value of carbon stored in the vegetation and soil of Hong Kong country parks, because the water bodies in Hong Kong are relatively small. Data on the total amount of carbon stored come from Delang and Yu (2009) who estimated the carbon sequestered in 1978, 1991, 1997, and 2004 (the data are extrapolated for the other years). The economic value of the carbon is based on the average price of US$20 per tonne of carbon dioxide in 2009, from the European Union Greenhouse Gas Emission Trading System (EU ETS). This figure is adjusted for inflation for the other years. Hence, the *Value of carbon sequestration* equals to Carbon stored × price of carbon dioxide. Because of lack of data, we do not estimate the *Value of carbon sequestration* for Singapore.

Cost of lost wetland (−)

Wetlands provide extensive ecological services, ranging from nutrient cycling to flood control, and from water purification to providing habitats for migratory birds and aquatic life, and to attractive aesthetic values (Russi *et al.*, 2013; Peh *et al.*, 2014). In the past, the critical importance of wetland functions was not well understood. Even though the understanding of the ecological importance of wetland has improved, the draining and filling of wetlands for land development continues. The reclamation of wetlands increases the GDP because of the high value of new land uses (for example for housing) (Wen *et al.*, 2007). Although the GPI takes into account the (positive) values of the new land use, it also includes the reduction in welfare that results from a loss of wetlands. As with the cost of lost farmland, some studies (e.g. Berik and Gaddis, 2011) name this cost the 'option value of wetlands'. The argument against using the term 'option value' is similar to that for lost farmland – calling this cost an option value would overstate overall GPI, as option values are psychic incomes.

Lawn (2008a), Forgie *et al.* (2008), and Makino (2008) accumulate the total costs for the ten years following the loss of wetlands, to reflect the non-substitutability of its ecological services by man-made capital. Wen *et al.* (2008), and Clarke and Shaw (2008) do not include the cost of lost wetlands. Here, we accumulate the *Cost of lost wetland cumulatively* over ten years. We estimate the value of wetlands using Costanza *et al.*'s (1997) data of US\$14,785 per hectare per year. The *Cost of lost wetland* equals the total number of hectares of wetland lost × value of wetland + cumulative cost of past ten years.

Comparisons with other studies

While most GPI studies use the indices introduced above, some studies contain a few variations. This is the case, for example, of the Wen *et al.* (2008) study of China, Lawn (2008a) study of Australia, and Clarke and Shaw (2008) study of Thailand. Wen *et al.* (2008) included the cost of excessive water use for irrigation and the cost of timber depletion, while they did not include the cost of underemployment, overwork, and lost wetland, because of insufficient data that could be used to estimate these costs. The cost of excessive water use for irrigation was included to reflect China's dire shortage of water and gross inefficiencies in water use, specially in irrigation (Wen *et al.*, 2008). A similar rationale was used for including the cost of timber depletion. From the 1950s to 1998 timber stocks in China were in decline, as the rate of timber harvesting exceeded the rate of forest regeneration (Wang and Delang, 2011; Delang and Wang, 2012, 2013).

Lawn (2008a) used the same items listed above, but added an Ecosystem Health Index, which is unique to his study. This item weighted the sum total of all environmental costs. This was an attempt to account for the impacts, besides the damage already done to the source and sink functions of the natural environment, of many resource-extractive and polluting activities, potentially extending to the general degradation of the ecosystem (Lawn, 2008a). An example given by Lawn (2008a) was strip mining, which involved both non-renewable resource depletion and the removal of terrestrial fauna and flora.

Clarke and Shaw's (2008) estimate of the GPI of Thailand added two categories (spiritual and political items) to the usual three (economic, social, environmental) (Table 4.2). The spiritual category includes the value of non-paid household labour, which is usually found in the social category, and the cost of commercial sex work. The political category included the welfare derived from public expenditure on roads and highways, and the cost of corruption. Other new items included the cost of urbanization, and the contributions to welfare of several forms of public expenditure.

Clarke and Shaw (2008) put the value of non-paid household labour into the spiritual category because they considered the modernization of the Thai economy since 1975 to have reduced the incentive for Thais to undertake substantial homework, a change which they considered damaging to the integrity of the family unit. The cost of commercial sex work was included because of its importance in Thailand, and because it was officially recognized by the government as contributing to the tourism industry (Clarke and Shaw, 2008). The cost reflected the poor conditions of sex workers, as many were 'forced to endure slave-like conditions', the exploitation of children and ethnic minorities, and the societal consequences of infidelity and the spread of sexually transmitted diseases (Clarke and Shaw, 2008). The cost of corruption was included to reflect the welfare cost of corruption-related activities by bureaucrats (including the military and police) and politicians (Clarke and Shaw, 2008). Corruption is a problem for many developing countries, which often lack the means to combat it. In Thailand, the problem of corruption has been worsening with the rapid economic growth that the country experienced during the last decades, since salaries of civil servants do not keep pace with the rising salaries in the private sector (Clarke and Shaw, 2008).

Table 4.2 GPI items used in the Thailand study

Items (brackets indicate whether it contributes to, or subtract from, welfare)

Economic	Spiritual	Political	Social	Environment
Consumption expenditure (+)	Value of non-paid household labour (+)	Welfare from public expenditure on roads and highways (+)	Welfare from public expenditure on health (+)	Cost of timber depletion (−)
Expenditure on consumer durables (−)	Cost of commercial sex work (−)	Cost of corruption (−)	Welfare from public expenditure on education (+)	Cost of land degradation (−)
Service from consumer durables (+)		Change in net foreign debt (+/−)	Cost of commuting (−)	Cost of urban waste-water pollution (−)
Distribution index (+/−)			Cost of urbanization (−)	Cost of air pollution (−)
Welfare from publicly-provided infrastructure (+)			Cost of noise pollution (−)	Cost of long-term environmental damage (−)

Source: Clarke and Shaw (2008)

These changes call for the importance of adapting the GPI to the conditions of different countries. Developing countries in particular will have different characteristics, a different economic structure (for example a greater reliance on social networks rather than formal arrangements to take care of elderly and children) which necessitates the adaptation of the GPI. In the case of Thailand, for example, the inclusion of the item *Welfare from public expenditures on health, Welfare from public expenditures on education*, and *Welfare from public expenditures on roads and highways* are particularly important. Improving education in particular is a form of investment in human capital that could yield benefits in the form of lower social costs, and will assist in future economic development. The impact of additional education expenditure on welfare in developing countries has been described as greater than that in developed countries (Clarke and Shaw, 2008).

Conclusions

This chapter has reviewed the different indicators used. There is considerable uniformity across studies. However, we have also underlined the need to adapt the items to the particular conditions of countries, because of the availability of data, and the importance of particular items to the characteristics of the country and its level of development. We have also reviewed the methods used to estimate the economic values of such indicators. As in the case of the choice of the items, the method to estimate their economic value is not uniform across studies. Methods may change because of differences in the availability of data, or different opinions by researchers as to which approach to use. Such lack of uniformity is one of the criticisms levied against the GPI, but we believe that it can also be considered an advantage because of the flexibility it bestows. Unlike with the GDP, with the GPI researchers can select the most useful items to assess the welfare of a country's population, and the best method to quantify the value of each item, to reflect the characteristics of the country under consideration. We contend that the flexibility offered by the GPI is a strength of the indicator, and not a weakness. In the following two chapters we present the results of the GPI of Hong Kong (Chapter 5) and the GPI of Singapore (Chapter 6).

Notes

1   Hong Kong is of course not a city-state, but a Special Administrative Region (SAR) of the People's Republic of China. However, it retains considerable independence in economic, financial, and political matters, so for the sake of this study we call a 'city-state'.
2   PM_{10} are particulate matter smaller than about 10 micrometers, which can penetrate the deepest part of the lungs such as the bronchioles or alveoli.

References

Berik, G. and Gaddis, E. (2011). The Utah Genuine Progress Indicator, 1990 to 2007: A Report to the People of Utah. In: www.utahpop.org/gpi.html

Bougheas, S. and Georgellis, Y. (1999). The effect of divorce costs on marriage formation and dissolution. *Journal of Population Economics*, *12*(3), 489–498.

Brown, S., Roberts, J. and Taylor, K. (2011). The gender reservation wage gap: evidence from British panel data. *Economics Letters*, *113*(1), 88–91.

Chou, K. L. and Chi, I. (2002). Financial strain and life satisfaction in Hong Kong elderly Chinese: Moderating effect of life management strategies including selection, optimization, and compensation. *Aging & mental health*, *6*(2), 172–177.

Clarke, M. and Shaw, J. (2008). Genuine progress in Thailand: a systems-analysis approach. In: Lawn, P. and Clarke, M. (Eds). *Sustainable Welfare in the Asia-Pacific: Studies using the Genuine Progress Indicator* (pp. 260–298). Cheltenham: Edward Elgar Publishing Ltd.

Cobb, C., Halstead, T. and Rowe, J. (1995). If the GDP is up, why is America down? *Atlantic-Boston*, *276*, 59–79.

Cobb, C. W. and Cobb, J. B. (1994). *The Green National Product: a Proposed Index of Sustainable Economic Welfare*. Lanham: University Press of America.

Costanza, R., d'Arge, R., de Groot, R., Farber, S., Grasso, M., Hannor, B., Limburg, K., Naeem, S., O'Neill, R., Paruelo, J., Rasins, R., Sutton, P. and van den Belt, M. (1997). The Value of the world's ecosystem services and natural capital. *Nature*, *387*, 253–260.

Daly, H., Cobb, H. E. and Cobb, J. B. (1989). *For the Common Good*. Boston: Beacon Press.

Delang, C. O. and Wang, W. (2012). 'Chinese Forest Policies in the Age of Decentralization (1978–1998)', *International Forestry Review*, *14*(1), 13–26.

Delang, C. O. and Wang, W. (2013). 'Chinese forest policy reforms after 1998: The case of the Natural Forest Protection Program and Slope Land Conversion Program', *International Forestry Review*, *15*(3), 290–304.

Delang, C. O. and Yu, Y. H. (2009). Remote Sensing-Based Estimation of Carbon Sequestration in Hong Kong. *Open Environmental Sciences*, *3*, 97–115.

El Serafy, S. (1989). The proper calculation of income from depletable natural resources. In: Ahmad, Y., El Serafy, S., Lutz, E. (Eds), *Environmental Accounting for Sustainable Development* (pp. 10–18). Washington DC: World Bank.

Fann, N. and Wesson, K. (2009). Estimating local health impacts using fine-scale air quality estimates and baseline incidence rates. *Epidemiology*, *20*(6), S36.

Forgie, V., McDonald, G., Zhang, Y., Patterson, M. and Hardy, D. (2008). Calculating the New Zealand Genuine Progress Indicator. In: Lawn, P. and Clarke, M. (Eds), *Sustainable Welfare in the Asia-Pacific: Studies using the Genuine Progress Indicator* (126–152). Cheltenham: Edward Elgar Publishing Ltd.

Fourie, J. (2006). Economic infrastructure: a review of definitions, theory and empirics. *South African Journal of Economics*, *74*(3), 530–556.

Giannelli, G. C., Mangiavacchi, L. and Piccoli, L. (2012). GDP and the value of family caretaking: how much does Europe care? *Applied Economics*, *44*(16), 2111–2131.

Gibbons, D. C. (2013). *The Economic Value of Water*. London: Routledge.

Golden, L. and Wiens-Tuers, B. (2008). Overtime work and wellbeing at home. *Review of Social Economy*, *66*(1), 25–49.

Hamilton, C. (1999). The genuine progress indicator methodological development and results from Australia. *Ecological Economics*, *30*, 13–28.

HKU. (2008). Work life balance in Hong Kong survey results. *Hong Kong University*. In: www.hku.hk/press/news_detail_5755.html. Retrieved on: 1 December 2011.

Ho, P. S. Y. (2007). Eternal mothers or flexible housewives? Middle-aged Chinese married women in Hong Kong. *Sex roles, 57,* 249–265.

Hyslop, N. P. (2009). Impaired visibility: the air pollution people see. *Atmospheric Environment, 43*(1), 182–195.

Jackson, T. and Styme, S. (1996). Sustainable Economic Welfare in Sweden: A pilot index 1950–1992. Stockholm: Stockholm Environment Institute.

Lawn, P. (2003). A theoretical foundation to support the Index of Sustainable Economic Welfare, Genuine Progress Indicator, and other related indexes. *Ecological Economics, 44,* 105–118.

Lawn, P. (2004). To operate sustainably or not to operate sustainability? – That is the long-run question. *Futures, 36,* 1–22.

Lawn, P. (2005). An assessment of the valuation methods used to calculate the Index of Sustainable Economic Welfare (ISEW), Genuine Progress Indicator (GPI), and Sustainable Net Benefit Index (SNBI). *Environment, Development and Sustainability, 7,* 185–208.

Lawn, P. (2006). An assessment of alternative measures of sustainable economic welfare. In: *Sustainable Development Indicators in Ecological Economics* (Chapter 7, pp. 139–165). Northampton, MA: Edward Elgar.

Lawn, P. (2008a). Genuine progress in Australia: time to rethink the growth objective. In: Lawn, P. and Clarke, M. (Eds). *Sustainable Welfare in the Asia-Pacific: Studies using the Genuine Progress Indicator* (pp. 91–125). Cheltenham: Edward Elgar Publishing Ltd.

Lawn, P. (2008b). Genuine progress in India: some further growth needed in the immediate future but population stabilization needed immediately. In: Lawn, P. and Clarke, M. (Eds), *Sustainable Welfare in the Asia-Pacific: Studies using the Genuine Progress Indicator* (pp. 191–227). Cheltenham: Edward Elgar Publishing Ltd.

Lawn, P. (2010). Facilitating the transition to a steady-state economy: Some macroeconomic fundamentals. *Ecological Economics, 69,* 931–936.

Lawn, P. and Clarke, M. (2008a). An Introduction to the Asia-Pacific region. In: Lawn, P. and Clarke, M. (Eds), *Sustainable Welfare in the Asia-Pacific: Studies using the Genuine Progress Indicator* (pp. 3–34). Cheltenham: Edward Elgar Publishing Ltd.

Lawn, P. and Clarke, M. (2008b). Why is Gross Domestic Product an inadequate indicator of sustainable welfare. In: Lawn, P. and Clarke, M. (Eds), *Sustainable Welfare in the Asia-Pacific: Studies using the Genuine Progress Indicator* (pp. 35–46), Cheltenham: Edward Elgar Publishing Ltd.

Lawn, P. and Clarke, M. (2008c). What is the Genuine Progress Indicator and how is it typically calculated. In: Lawn, P. and Clarke, M. (Eds), *Sustainable Welfare in the Asia-Pacific: Studies using the Genuine Progress Indicator* (pp. 47–68), Cheltenham: Edward Elgar Publishing Ltd.

Lawn, P. and Clarke, M. (2008d). In defence of the Genuine Progress Indicator. In: Lawn, P. and Clarke, M. (Eds), *Sustainable Welfare in the Asia-Pacific: Studies using the Genuine Progress Indicator* (pp. 69–88), Cheltenham: Edward Elgar Publishing Ltd.

Lawn, P. and Clarke, M. (2008e). Genuine progress across the Asia-Pacific region: comparisons, trends, and policy implications. In: Lawn, P. and Clarke, M. (Eds), *Sustainable Welfare in the Asia-Pacific: Studies using the Genuine Progress Indicator* (pp. 333–361). Cheltenham: Edward Elgar Publishing Ltd.

Lawn, P. and Clarke, M. (2010). The end of economic growth? A contracting threshold hypothesis. *Ecological Economics, 69*, 2213–2223.

Lu, J. R. and Peng, C. (1999). The trend state and life-time of urban residents in China. *China Soft Science, 6*, 90–93.

Makino, M. (2008). Genuine progress in Japan and the need for an open economy GPI. In: Lawn, P. and Clarke, M. (Eds), *Sustainable Welfare in the Asia-Pacific: Studies using the Genuine Progress Indicator* (pp. 153–190). Cheltenham: Edward Elgar Publishing Ltd.

Martinet, V. (2012). *Economic Theory and Sustainable Development: What can we preserve for future generations?* London: Routledge.

McCollister, K. E., French, M. T. and Fang, H. (2010). The cost of crime to society: New crime-specific estimates for policy and program evaluation. *Drug and alcohol dependence, 108*(1), 98–109.

Mjelde-Mossey, L. A., Chi, D. S. W. and Chow, N. (2002). Volunteering in the social services: Preferences, expectations, barriers, and motivation of aging Chinese professionals in Hong Kong. *Hallym International Journal of Aging, 4*(1), 31–44.

Neumayer, E. (1999). The ISEW – Not an index of sustainable economic welfare. *Social Indicators Research, 48*, 77–101.

Neumayer, E. (2000). On the methodology of the ISEW, GPI, and related measures: Some constructive suggestions and some doubt on the threshold hypothesis. *Ecological Economics, 34*, 347–361.

Nguyet Hong, V. X., Clarke, M. and Lawn, P. (2008). Genuine progress in Vietnam: the impact of the Doi Moi reforms. In: Lawn, P. and Clarke, M. (Eds), *Sustainable Welfare in the Asia-Pacific: Studies using the Genuine Progress Indicator* (pp. 299–330). Cheltenham: Edward Elgar Publishing Ltd.

Peh, K. S. H., Balmford, A., Field, R. H., Lamb, A., Birch, J. C., Bradbury, R. B. and Hughes, F. M. (2014). Benefits and costs of ecological restoration: Rapid assessment of changing ecosystem service values at a UK wetland. *Ecology and evolution, 4*(20), 3875–3886.

Postel, S. and Richter, B. (2003). *Rivers for Life: Managing water for people and nature.* Island Press.

Quah, E. and Boon, T. L. (2003). The Economic Cost of Particulate Air Pollution on Health in Singapore. *Journal of Asian Economics, 14*, 73–90.

Raaschou-Nielsen, O. *et al.* (2013) Air pollution and lung cancer incidence in 17 European cohorts: prospective analyses from the European Study of Cohorts for Air Pollution Effects (ESCAPE), *The Lancet Oncology, 14*(9), 813–822.

Regus. (2011). Singapore Workers Work the Day AND Night Away. 23 November 2011. Retrieved 14 October 2013 from http://press.regus.com/singapore/singapore-workers-work-the-day-and-night-away

Russi, D., ten Brink, P., Farmer, A., Badura, T., Coates, D., Förster, J. and Davidson, N. (2013). *The economics of ecosystems and biodiversity for water and wetlands,* London and Brussels: IEEP.

Sajardo, A. and Serra, I. (2011). The economic value of volunteer work methodological analysis and application to Spain. *Nonprofit and Voluntary Sector Quarterly, 40*(5), 873–895.

Van Kooten, G. C., Johnston, C. and Xu, Z. (2012). Economics of forest carbon sequestration. Resource Economics and Policy Analysis (REPA) Research Group, Department of Economics, University of Victoria, 25.

Wang, W. and Delang, C. O. (2011). Chinese Forest Policies in the Age of Ideology (1949–1978), *International Forestry Review, 13*(4), 416–430.

Wen, Z. G. (2005). 'Capital extension methodology: the simulation study into policy alternatives towards sustainable development', Dissertation submitted to Tsinghua University in partial fulfillment of the requirement for the degree of Doctor of Engineering, Library of Tsinghua University.

Wen, Z., Yang, Y. and Lawn, P. (2008). From GDP to the GPI: quantifying thirty-five years of development in China. In: Lawn, P. and Clarke, M. (Eds). *Sustainable Welfare in the Asia-Pacific: Studies using the Genuine Progress Indicator* (pp. 228–259). Cheltenham: Edward Elgar Publishing Ltd.

Wen, Z., Zhang, K., Du, B., Li, Y. and Li, W. (2007). Case study on the use of genuine progress indicator to measure urban economic welfare in China. *Ecological Economics*, *63*, 463–475.

Wong, D. (1996). Housewife role and women's psychological well-being: The case of Tuen Mun. *Hong Kong Journal of Social Work*, *30(1)*, 48–56.

5 The Genuine Progress Indicator of Hong Kong: results

Introduction

In this chapter, we estimate each individual item of the GPI of Hong Kong, and then provide the aggregate figures. Apart from allowing us to understand better which aspects have been improving and which deteriorating, assessing the individual items of the GPI helps us identify those areas which need to be addressed. All data in the figures are per capita, in 2009 HK$.

In the period under consideration (1968 to 2010) Hong Kong has experienced considerable, sustained, economic (as measured by the GDP) growth. We provide a brief description of the economic changes that occurred in Hong Kong (and Singapore) in Chapter 7, but two events should be kept in mind when analysing the data presented in this chapter. First, the slump that occurred in 1997, because of the Asian Financial Crisis, followed by the SARS epidemic. Second, the gradual deindustrialization of Hong Kong, and investment in the Chinese province of Guangdong, that started with China's 'Open Door Policy' from 1978. Between 1978 and 1997, trade between Hong Kong and China grew at an average rate of 28 per cent per annum, and there was a major shift of labour-intensive industries from Hong Kong to the mainland, to take advantage of its cheap labour. This also brought about the transformation of Hong Kong's economy, from one based on manufacturing to one based on services, as was evident from the increase of employment in the service sector from 52 per cent of the labour force in 1981 to 80 per cent in 2000, while the labour force involved in manufacturing dropped from 39 per cent to 10 per cent during the same period. These transformations of the Hong Kong economy help explain some of the trends we will be seeing in this chapter.

Economic items

Personal and public consumption expenditure (+)

The item that most influences the final GPI data is of course *Personal and public consumption expenditure*, which is also the most important component of the GDP (Figure 5.1). This item increased from HK$27,844 in 1968 to HK$172,686

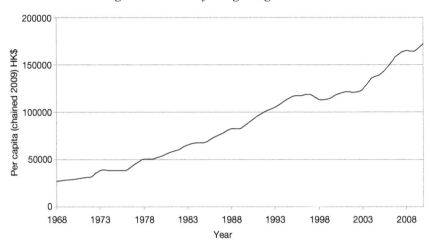

Figure 5.1 Personal and public consumption expenditure, Hong Kong

in 2010, representing an increase of about 4.44 per cent per year. However, the figure also shows the important drop that occurred following the 1997 Asian Financial Crisis. The Asian Financial Crisis had an impact on many branches of the economy, in particular the property sector (Lai *et al.*, 2006a, 2006b) and tourism (Song and Lin, 2009). In 1997, *Personal and public consumption expenditure* was at HK$119,139 per capita, and by 1999 it had dropped to HK$113,953. However, the manufacturing sector was relatively unaffected, since it was already based in the Pearl River Delta, and 'goods continued to be exported through Hong Kong normally' (Siu and Wong, 2004: 62). As a consequence, the economy rebounded relatively rapidly, and in 2001 it had recovered to the 1997 level. During the following years it kept increasing. The Global Financial Crisis of 2008 had a comparatively lower impact on the *Personal and public consumption expenditure* in Hong Kong.

Defensive and rehabilitative expenditure (−)

The second most important item (this one negative) is *Defensive and rehabilitative expenditure*, which includes those items that people need to prevent or cure illnesses, or otherwise do not increase welfare, or well-being (Figure 5.2). As with *Personal and public consumption expenditure*, this has been steadily increasing. It is interesting to note that from 1997 to 2003 it has been stable rather than dropping, as *Personal and public consumption expenditure* did. This item includes some of the expenditures people resist cutting, even in times of economic problems. While some of the items are expenditures (e.g. savings or recreation) that may be cut in times of crisis, others are percentage (in many cases 25 per cent) of the most basic expenditures (such as food, rent, and fuel), which many people may find difficult to cut, and others again may be expenses (e.g. on

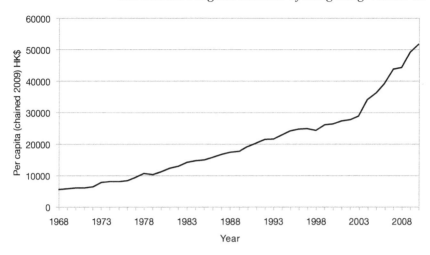

Figure 5.2 Defensive and rehabilitative expenditure, Hong Kong

education) which may even increase, as young people extend their study years to increase their job prospects (Thomas *et al.*, 2004; Gottret *et al.*, 2009; Meghir and Pistaferri, 2011). It is of course questionable whether some of the items included in *Defensive and rehabilitative expenditure* should indeed be considered costs that lower welfare. We chose to include the same items (as far as possible) that were included by other researchers, so as to facilitate comparison. A debate can (and should) take place as to what exactly one should include in this category (as reported in Chapter 4).

Expenditure on consumer durables, and services from consumer durables (– and +)

As mentioned in Chapter 4, *Expenditure on consumer durables* is subtracted from the GPI, because these items provide welfare for a number of years (Figure 5.3). These services are added to the GPI through the item *Services from consumer durables*, for the duration that these consumer products last, which we assume to be seven years. The welfare (the services) provided by these items throughout their lifetime are added, instead of the price of the product on the day of purchase. The difference between these two values is actually not very large, over the long timeframe we consider.

Weighted adjusted consumption expenditure

Figure 5.4 shows the *Weighted adjusted consumption expenditure*, together with its constituents (the indices that are subtracted to estimate the GPI are plotted as negative). The figure does not show the *Income distribution index*, which is used to weight the previous four items to obtain the *Weighted adjusted consumption expenditure*. Income inequality is measured using the Gini

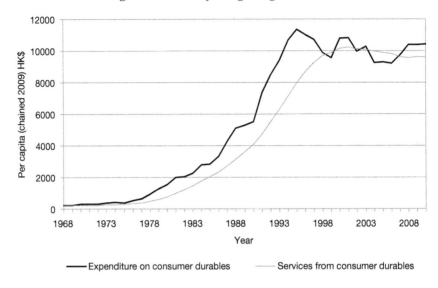

Figure 5.3 Expenditure on consumer durables and Services from consumer durables, Hong Kong

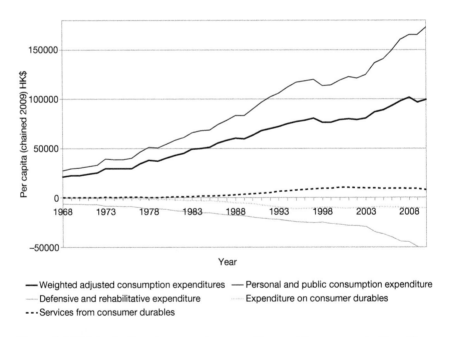

Figure 5.4 Weighted adjusted consumption expenditure and its components, Hong Kong

coefficient (Gini, 1921). In Hong Kong, the Gini coefficient has been increasing over time (Chan *et al.*, 2014), from 0.45 in 1981 to 0.53 in 2006, where 0 represents perfect equality (everybody has the same), and 1 represents perfect inequality (everything belongs to one person). Hong Kong's Gini coefficient is higher than that of all countries that belong to the Organisation of Economic Cooperation and Development (OECD), which means that it has very high inequality for an industrial country (World Bank, 2014). However, one should bear in mind that the Gini coefficient measures *income* inequality, and does not take into consideration subsidies that may exist for the lower-income households. In 2011, in Hong Kong 29.1 per cent of the population was living in public rental housing and 17.1 subsidized home ownership housing (Lui, 2013). Both kind of housing cater for the households with the lowest aggregate incomes, and provide very cheap housing in a city where housing prices are among the highest in the world (Craig, 2011). For this reason, in the case of Hong Kong the Gini coefficient figures are slightly misleading. Unfortunately we do not have other figures on inequality for Hong Kong.

As we can see, the *Weighted adjusted consumption expenditure* has been increasing steadily over the period under consideration. In 1968, it was of HK$21,335 per capita, while in 2010 it was of HK$99,561 per capita, an increase of 3.7 per cent a year. However, this increase was far below that of *Personal and public consumption expenditure*, in particular because of the increasing importance of *Defensive and rehabilitative expenditure*.

Services yielded from fixed capital (+)

Services yielded from fixed capital have steadily increased over the decades under consideration (Figure 5.5). We can also see that the financial crisis has not affected government expenditure: Hong Kong government expenditure on infrastructure and public work is notoriously high, and disregards economic and financial cycles. Indeed, it may be argued that the government has been using the synergies between public and private investment to spur economic growth (Wang, 2002). These synergies become more important during economic crises, when governments use the externalities of public infrastructure projects to spur private production growth (Wang, 2002), although in the case of Hong Kong public infrastructure may have lower externalities because the manufactory base has moved to Guangdong province. As discussed in Chapter 4, we multiply Gross domestic fixed capital formation by 0.1 and assume that the investment generates benefits for ten years (many would last much longer). The curve shows the smoothing effect of our calculation method.

Change in net foreign assets (+/−)

Change in net foreign assets reflects the large foreign reserves held, and investments that Hong Kong businesses and government agencies undertake, outside Hong Kong, in particular in the neighouring Pearl River Delta in China,

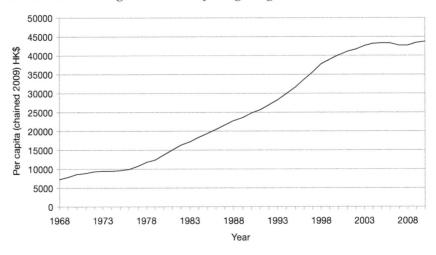

Figure 5.5 Services yielded from fixed capital, Hong Kong

and increasingly beyond the Pearl River Delta (Liao and Chan, 2011) (Figure 5.6). The figure shows the drop in 1997 due to the Asian Financial Crisis, when the regional economies slowed down. The figure also shows that while until 1997 net foreign investments of Hong Kong was relatively small, as Hong Kong became a Special Administrative Region of China on 1 July 1997, foreign investment increased considerably (China is considered a different country in these statistics). We are unable to explain the drop from 2002 to 2005, though it rapidly recovered afterwards, and has stabilized just under HK$40,000 per capita (in chained 2009 HK$). It is worth noting that data for foreign assets were absent prior to 1991 and the data used for this study were extrapolated through an exponential trend line.

Economic sub-index

Figure 5.7 shows all the economic items (except the aggregated *Weighted adjusted consumption expenditure*). As we can see, *Personal and public consumption expenditure* is the most important item, by far outweighing all others. It is also the only item that dropped slightly during the Asian financial crisis. The economic items of the GPI will be weighted against the social and environmental indices.

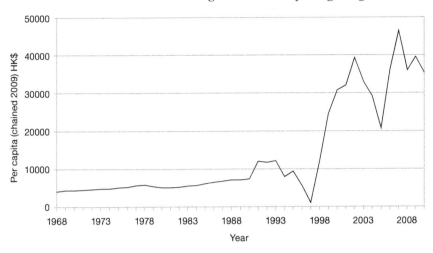

Figure 5.6 Change in net foreign assets, Hong Kong

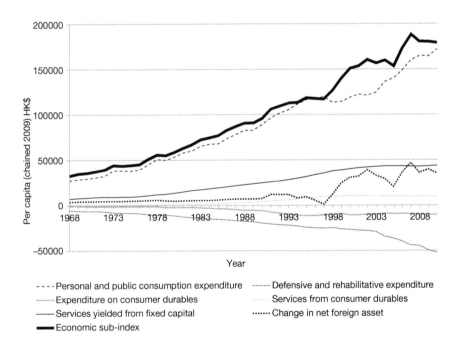

Figure 5.7 Economic items of the GPI, Hong Kong

Social items

We turn to the social items, and present again each item independently, before comparing them in the last figure of this section.

Value of non-paid household labour (+)

The *Value of non-paid household labour* has been increasing steadily during the 43 years under consideration, from HK$5,387 in 1968 to HK$13,416 in 2010, an average growth rate of 2.2 per cent per annum (Figure 5.8). According to Wong (1996) and Ho (2007), the attitude towards housewives has drastically changed over the past four decades, as their status has changed from one of appreciation to one with an increasing feeling of stigmatization. The trend towards more women working for a wage is further amplified by the increasing availability of lowly paid domestic helpers from Indonesia and the Philippines to do household chores. In 2013, there were some 320,000 foreign domestic helpers in Hong Kong, 50 per cent of which were from the Philippines, and 47 per cent from Indonesia, earning a salary of only HK$4,010 a month, well below the minimum wage (SCMP, 2014). The rise in value of non-paid household labour should be ascribed to the increase in average monthly wages (used to estimate the value of this item) throughout the period under consideration, rather than an increase in the number of people engaging in non-paid household labour.

Value of volunteer labour (+)

The *Value of volunteer labour* has remained very low during these years, although it has increased considerably over the last ten years (Figure 5.9). In 1968, the

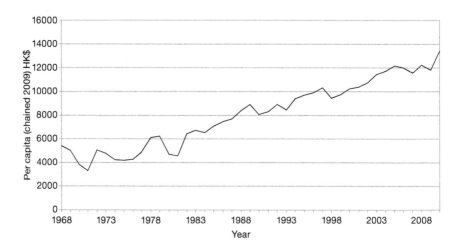

Figure 5.8 Value of non-paid household labour, Hong Kong

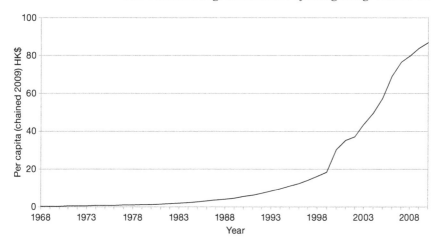

Figure 5.9 Value of volunteer labour, Hong Kong

value was of HK$0.27 per capita, and in 2010 it grew to HK$87, still a very low number, although it represents a yearly increase of almost 15 per cent. To estimate these values we use official government statistics from the Social Welfare Department of Hong Kong to estimate the number of people who engage in volunteer labour, and the minimum wage of HK$28 in 2010, discounted for inflation for the previous years. The low values reported in Figure 5.9 are very likely due to the fact that not all volunteer labour is recorded in the official statistics. The upward trend in registered volunteer labour is an encouraging sign, although it is uncertain what would be the shape of the trend if unregistered volunteer labour was included.

Cost of crime (−)

The *Cost of crime* is relatively low, and has remained low over the period under consideration, reflecting the low crime rate of Hong Kong (lower than that of Japan and Singapore, known to be very safe countries) (Figure 5.10). A 2010 study based on a 2006 dataset shows that compared to 33 capitals and main cities throughout the world, Hong Kong recorded lower rates of victimization for most types of crime: the combined victimization rate for the ten most common crimes was about three times lower than the international average (7.8 per cent compared to 21.5 per cent). Similarly, only one third as many respondents in Hong Kong compared to 22 main cities in the developed world said that they had been exposed to drug-related problems. On the other hand, Hong Kong has had high levels of consumer fraud, nearly twice as many as the international average for 25 cities in developed countries (Broadhurst *et al.*, 2010). The per capita cost of crime stood relatively stable over the study period, slightly increasing from HK$1,710 in 1968 to HK$1,872 in 2010, with a drop in between.

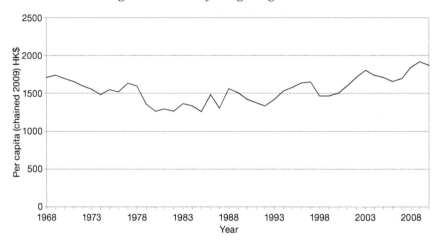

Figure 5.10 Cost of crime, Hong Kong

Cost of unemployment and underemployment

Hong Kong's overall unemployment rate has been relatively low, although as an open financial centre it is exposed to global economic changes, and was hit hard by the economic crisis of 2008. As a consequence, the unemployment rate increased from 4.4 per cent in the first quarter of 2007 to 5.3 per cent in the second quarter of 2009, with layoffs occurring mostly in the financial and other economy-sensitive sectors (Lee *at al.*, 2010). By 2011, the unemployment rate had dropped to 3.4 per cent (Hong Kong Census and Statistics Department, 2012), although there were considerable regional variations: according to the 2011 Yuen Long Detailed Census, approximately 11.2 per cent of residents in the Yuen Long district were unemployed (YLDC, 2011; Bouffard *et al.*, 2013). While the unemployment rate is relatively low, one should remember that 'Asian societies […] are not used to high unemployment rates and have limited welfare provisions' (Lee *et al.*, 2010: 131).

Figure 5.11 shows that the *Cost of unemployment and underemployment* was relatively low until 1997, but has since been rising, especially from 1998 to 2003. Over the whole period, the per capita *Cost of unemployment and underemployment* rose from HK$892 in 1968 to HK$4,592 in 2010, representing an average growth rate of 4 per cent per annum. The fact that we use the average annual income to estimate this item contributes to the rising costs, since the average annual income has been rising during the period under consideration, even adjusted for inflation.

Cost of overwork (−)

The *Cost of overwork* (Figure 5.12) has been increasing steadily over the years, reflecting an increasing demand on labour, as well as the culture of hard work that

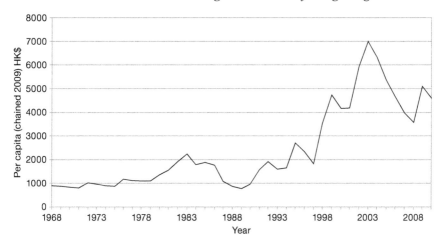

Figure 5.11 Cost of unemployment and underemployment, Hong Kong

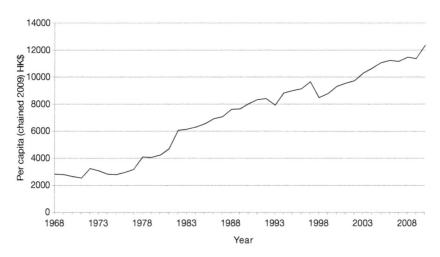

Figure 5.12 Cost of overwork, Hong Kong

exists in Hong Kong (Leung and Chan, 2012). The cost of overwork or lost leisure time rose from HK$2,844 in 1968 to HK$12,361 in 2010, representing an average growth rate of 3.6 per cent per annum. This is the largest negative item in the Social sub-index throughout the study period – fluctuating between 50 per cent and 80 per cent of the total. The overwork culture has become somewhat of a dilemma in Hong Kong, where extended working hours, which are more often than not unpaid, are seen as evidence of commitment and responsibility towards the company, as well as a positive trait that portrays a hardworking attitude, though with significant negative social impacts, such as reduced leisure and family time (Carney, 2011).

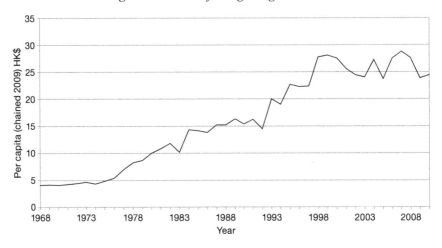

Figure 5.13 Cost of family breakdown, Hong Kong

Cost of family breakdown (−)

The *Cost of family breakdown* is the lowest of all Social sub-index items, and contributes to less than 1 per cent of the total, throughout the study period (Figure 5.13). Hong Kong, as a Chinese society that upholds traditional family values, sees divorces negatively. However, the value of this index is increasing by an average of 4.3 per cent per annum (from HK$4 in 1968 to HK$28 in 1998, stabilizing after that), which means that divorce cases have been increasing. A combination of factors contribute to the increase in divorce, including a weakening of family values, divorces becoming more socially acceptable, increasing overwork (which results in reduced family interactions), or unemployment and underemployment (which results in more family conflicts) (Hung *et al.*, 2004; Sullivan, 2005).

Direct disamenity of water and air pollution (−)

This item is estimated using capital expenditure on air and water related projects. The *Direct disamenity of air pollution* and the *Direct disamenity of waste-water pollution* are small items, with per capita costs of only HK$113 and HK$135 respectively in 2010 (Figure 5.14). The low amounts reported in Figure 5.14 are not due to low levels of air and waste-water pollution, but to the low investments by the government to address this problem. The Hong Kong government tends to either deny the problem, or argue that the pollution is blown from the Guangdong province in China, and therefore there is little it can do (Delang and Cheng, 2013):

> In fact, the air is not all that bad. In fact, the air this year is better than it was last year, and last year was better than the year before. The air quality today

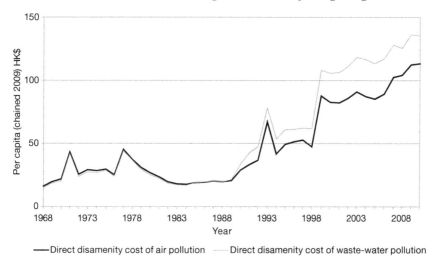

Figure 5.14 Direct disamenity of air pollution and of water pollution, Hong Kong

is not inferior to Washington DC, if I may say so. But I'm really not complacent, and I know there is a lot of work to do. Not only in Hong Kong – in Hong Kong we have limitations on what we can do. We have now cleaned up our old vehicular fleet. We have two power stations to look at, and we have to make sure they are up to the mark. But beyond that, it is all outside of Hong Kong. And on that we are working very hard with the mainland, particularly in Guandong Province.

(Donald Tsang, former Chief Executive of Hong Kong,
as reported in the Hong Kong Journal, 2006)

In spite of the little expenditure, the value of these items increased by 4.9 per cent per annum for air pollution and 5.2 per cent per annum for waste-water pollution during the period under consideration. The increased investment in air and water pollution abatement is expected, as the economy grows and Hong Kong becomes more populated (Thach *et al.*, 2010). However, equally important is the fact that as people become wealthier and better educated, including about the negative impact of pollution, their expectations of the state of the environment rises (DeGolyer, 2008). At the same time, greater wealth translates in higher costs of air pollution, as the loss of GDP gradually increases if people are unable to work, and as businesses leave Hong Kong because of the high pollution (WSJ, 2012).

Social sub-index

Figure 5.15 shows the social items of the Hong Kong GPI, with their respective sign. Only two items have a positive sign, the *Value of non-paid household*

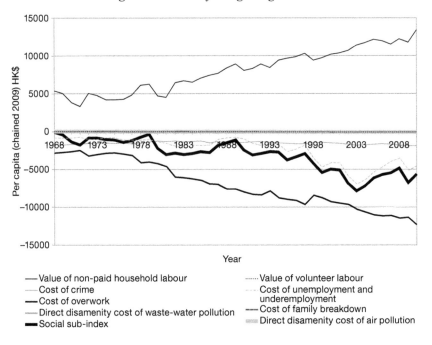

Figure 5.15 Social items of the Hong Kong GPI

labour, and the *Value of volunteer labour*. The majority have a negative sign, which reflects the choice of the items, but remind us also of the social costs that accompany economic activities. The value of all items is increasing with time, which reveals the fact that as the economy is growing the social benefits and costs increase. Overall, the social items are relatively small, which means that the social index will have little impact on the GPI.

Environmental items

We now turn to the environmental items. As in the previous sections, we present first each index, and then the aggregated sub-index. Almost all environmental items have a negative sign, since economic activities consist predominantly in the extraction, transformation, use and disposal of natural resources, all of which have a negative impact on the environment.

Cost of non-renewable resource depletion (−)

We start with the *Cost of non-renewable resource depletion*, which for Hong Kong mainly consist in quarried minerals (Figure 5.16). Despite its small size, Hong Kong has a relatively large number of mineral deposits, some of which were exploited commercially. In particular, lead was mined in the Lin Ma Hang

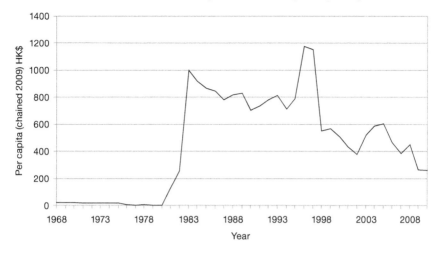

Figure 5.16 Cost of non-renewable resource depletion, Hong Kong

area from the 1910s on and off until 1958, wolframite was mined at the Needle Hill Tungsten Mine from the 1930s to 1967, iron was mined in Ma On Shan from 1906 to 1976, and graphite was mined in the West Brother Island (Tai Mo To) from 1952 to 1973 (CEDD, 2014). Increasing costs, especially labour costs, and declining ore prices led to a closure of mining operations in Hong Kong, and now minerals are no longer mined. The *Cost of non-renewable resource depletion* was particularly high in the 1980s and 1990s, after mining operations were closed and the area rehabilitated, but has since dropped considerably.

Cost of agricultural land degradation (−)

The *Cost of agricultural land degradation* was relatively high at the beginning of the period under consideration, at HK$45 in 1968, but rapidly declined. During the years under consideration there has been a transformation of the farming sector. In the late 1960s and early 1970s much of the land was still farmed in Hong Kong, including rice. Nowadays, these farms have been replaced by small but intensive vegetable and livestock production. In 2013 there were about 2,400 farms in the territory, employing directly about 4,400 farmers and workers. Out of a total of 110,000 ha of Hong Kong land surface, 298 ha (0.27 per cent of the total territory) are used to grow vegetables, 276 ha for orchards, 137 ha for flowers, and 18 ha for field crops. In spite of the very small land surface used for agriculture, local production still plays a role. While only 2 per cent of fresh vegetables consumed in Hong Kong are produced locally, 59.6 per cent of live poultry and 6.8 per cent of live pigs consumed locally are produced in the territory (AFCD, 2014a). The *Cost of agricultural land degradation* shows the decline in the agricultural output in Hong Kong. In 2010, this value had dropped to HK$2 per person (Figure 5.17).

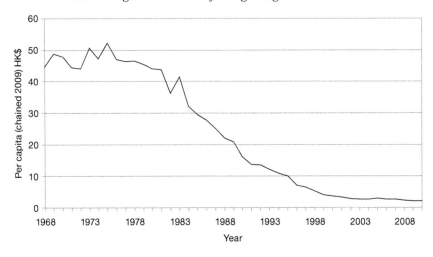

Figure 5.17 Cost of agricultural land degradation, Hong Kong

Cost of fisheries depletion (−)

The *Cost of fisheries depletion* is calculated based on the value of the catch (the larger the catch, the higher the cost) and the price of the fish, on the assumption that the more fish are caught, the more should be invested to sustain the flow of income. From 1968 to 1979 the costs had an upward trend (which means that fish catch and/or prices increased during this period) (Figure 5.18). In the 1980s and 1990s the costs decreased, stabilizing in the 2000s. Since during this period the price of the fish increased considerably, the figure underestimates the drop in the amount of fish caught.

The *Cost of fisheries depletion* brings together capture fishery and aquaculture, which in Hong Kong only makes a very small contribution to total production. In 2013, the fishing industry produced an estimated 170,129 tonnes of produce, valued at about $2,338 million. The industry now consists of some 4,000 fishing vessels and some 8,800 local fishermen (AFCD, 2014b). Fishing activities are mainly conducted in the waters of the adjacent continental shelf in the South China Sea. Within, and close to, Hong Kong territorial waters, the main fishing method is trawling. Trawling is a very destructive fishing method, and is partly responsible for the drop in the amount of fish caught in the period under consideration (other causes being overfishing, pollution, and land reclamation, which worsen water quality). Only on 13 October 2010, several years after conservationists prompted the government to do so, did the Hong Kong government announce a ban to all bottom and mid-water trawling activities. The ban came into effect on 31 December 2012, with a HK$1.72 billion (US$219 million) trawler buyout scheme that provides ex gratia payments for affected inshore trawler owners and larger vessels that generally do not operate in Hong Kong waters (Badalamenti *et al.*, 2012; WWF, 2014). Aquaculture (which

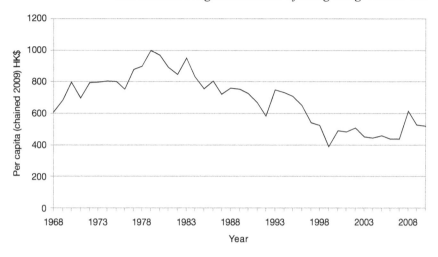

Figure 5.18 Cost of fisheries depletion, Hong Kong

includes marine fish culture, pond fish culture and oyster culture) makes a small contribution to the total amount of fish landing in Hong Kong. In 2013 production from the aquaculture sector was 3,300 tonnes, valued at $147 million, 2 per cent in weight and 6 per cent in value of the total fisheries production.

Cost of air pollution (−)

Hong Kong has very high rates of air pollution. The Hedley Environmental Index (http://hedleyindex.sph.hku.hk) estimated that there are over 3,000 premature deaths every year caused by air pollution, and over 200,000 hospital bed days a year. Sources of air pollution mainly comprise road transport and electricity generation. The government undertook some limited measures to address roadside emissions. Since Hong Kong is densely inhabited, much of the pollution generated is from vehicles in close proximity to densely populated areas. As such, improvements came in the form of replacing large diesel fleets with 'liquefied petroleum gas (LPG) engines, improving vehicle and fuel emission standards, emission reduction technology retrofit programmes, cleaner fuels, improved maintenance programmes, and increased emission checks and penalties for excessive dark smoke emissions' (Leverett *et al.*, 2007). One improvement was the (limited) reduction of the proportion of pre-Euro and Euro I diesel vehicles from 89 per cent in 1999 to 53 per cent in 2006 (Leverett *et al.*, 2007).

Yet, the actions by the government are surprisingly limited, considering the scale of the problem. From 1987 to 2014 the Hong Kong government had the same Air Pollution Index, with much lower standards than those recommended by the World Health Organisation (WHO). In January 2014, the government introduced much improved standards, the Air Quality Objectives and Air Quality Health Index (AQHI). The new AQHI monitors concentrations of four major pollutants on a three-hour moving average and alerts residents to the potential health risks posed by the air on a scale ranging from one (low health risk) to 10+ (serious health risk). Yet these still fall short of those recommended by the WHO. Table 5.1 compares the standards of the AQHI to those recommended by the WHO. Only for one pollutant (Nitrogen dioxide) are the standards the same. For all the others they are between 60 per cent higher (for Ozone) to 625 per cent higher (for Sulphur dioxide).[1]

Together with the new AQHI, the Environmental Protection Department launched an App on 30 December 2013, to display the AQHI and the concentrations of pollutants, including PM10 and PM2.5, recorded at 14 general and roadside monitoring stations (SCMP, 2013a). However, the government has been doing very little to actually reduce the level of pollution in Hong Kong. As James Middleton, chairman of city charity Clear the Air (a Hong Kong NGO advocating for cleaner air) assents, 'It's pointless having an index saying that you're going to die. What they should be doing is stopping the reasons that you're going to die' (TimeOut, 2014b).

Apart from roadside pollution generated by vehicles, a considerable amount of air pollution is blown from the Pear River Delta region. Despite the relocation of manufacturing industries from Hong Kong to mainland China, air and water pollution from these areas continue to affect Hong Kong (Leverett *et al.*, 2007). There have been joint agreements between the governments of Guangdong and Hong Kong to control pollution levels (ibid.). One such agreement, the Pearl River Delta Regional Air Quality Management Plan in 2003, imposed emission caps for sulphur dioxide, oxides of nitrogen, and respirable suspended particulates on power plants in the region, and also allocated them tradable emission credits (ibid.). There is a growing urgency to undertake new agreements, and enforce the

Table 5.1 Comparison of Hong Kong's AQO, AQHI, and the WHO's air quality guidelines (number of allowed exceedances noted in brackets)

Pollutant	Averaging time	1987–2014 AQOs ($\mu g/m^3$)	Starting from 2014 AQHIs ($\mu g/m^3$)	WHO AQGs ($\mu g/m^3$)
Sulphur dioxide (SO_2)	24-hours	350 (1)	125 (3)	20 (3)
Particulate matter (PM10)	24-hours	180 (1)	100 (9)	50 (3)
	1 year	55 (n/a)	50 (n/a)	20 (n/a)
Fine particulate matter (PM2.5)	24-hours	–	75 (9)	25 (9)
	1 year	–	35 (n/a)	10 (n/a)
Nitrogen dioxide (NO_2)	1-hour	300 (3)	200 (18)	200 (18)
Ozone (O_3)	8-hours	–	160 (9)	100 (9)

Source: EPD (2014)

existing ones, as the current lack of commitment from both sides means that the air quality of the region will only get worse (ibid.).

It is not surprising that the second most important environmental item is the *Cost of air pollution*, which increased steadily from HK$112 per capita in 1968 to HK$830 per capita 2010, at a rate of 5 per cent yearly, but mainly from the mid-1980s (Figure 5.19). The *Cost of air pollution* includes the value of lost productivity, which reflects the fact that air pollution causes premature deaths.

Cost of water pollution (−)

Figure 5.19 also shows the changes in *Cost of water pollution*, based on government capital expenditure to address waste-water pollution. Government statistics have indicated an overall improvement in water quality over the past four decades; for example, the decline in *Escherichia coli* (*E. coli*) levels in Hong Kong's harbours. This has had much to do with the implementation of the Strategic Sewage Disposal Scheme in 2001, which was later renamed the Harbour Area Treatment Scheme (HATS). The scheme had two stages that involved primary and secondary treatment of waste water. The commissioning of stage one of the scheme in 2001, along with the Tolo Harbour Effluent Export Scheme and the closure of Kai Tak Airport, resulted in significant improvements to the marine water quality in Victoria Harbour, as shown by its 99 per cent compliance with marine Water Quality Objectives (WQOs) (HKIE,

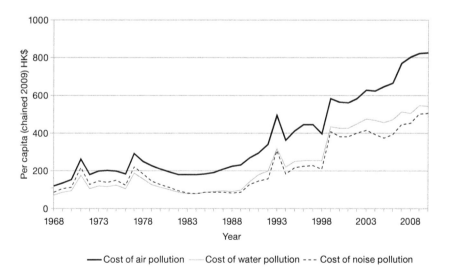

Figure 5.19 Cost of air pollution, water pollution, and noise pollution, Hong Kong

Note: The *Cost of air pollution, water pollution* and *noise pollution* come from official statistics from the Hong Kong Annual Digest of Statistics under the entry 'Capital expenditure devoted to environmental projects by type'. The shape of these three curves is similar, because they are only presented as a percentage of the total capital expenditures.

2009; Leverett *et al.*, 2007). In addition, the government implemented the Water Pollution Control Ordinance in all Water Control Zones in 1996. This was critical in reducing the amount of toxic metals and other harmful substances that were discharged into the harbour, even though highly contaminated residues remained, because of decades of accumulation (ibid.).

Another crucial element in waste-water control has been the implementation of the Livestock Waste Control Scheme and the gradual improvement of rural sewerage networks, which helped improve river water quality after 1997 (EPD, 2012; Leverett *et al.*, 2007). It was recorded that 82 per cent of river monitoring stations had good or excellent quality in 2005 (Leverett *et al.*, 2007). The government achieved considerable progress in improving water quality through a variety of schemes and ordinances. The areas that require further attention include sewerage treatments in rural areas, the long-term development of HATS, and dealing with the outflow of waste water from mainland China (ibid.).

Cost of noise pollution (−)

Hong Kong is a very densely-inhabited region. It has a total land mass of 1,104 sq. km, but the whole population is concentrated on 24 per cent of the land. This 24 per cent includes housing (41 sq. km), rural settlements (35 sq. km) roads (40 sq. km), and open spaces, such as urban parks (25 sq. km). Areas not included in the 24 per cent of built areas include agricultural land (51 sq. km) and fish ponds (17 sq. km) (Planning Department, 2013). Approximately 40 per cent of Hong Kong territory is covered by Country Parks. The high concentration increases noise pollution because housing is often built at close proximity to roads, including highways, which means that much money is spent on noise barriers. The *Cost of noise pollution* (Figure 5.19) is estimated based on government capital expenditure for noise abatement programmes, mainly noise barriers.

Cost of solid waste (−)

The disposal of solid waste is a topic of concern in Hong Kong, as land for landfills has to be claimed from Country Parks. Hong Kong has three strategic landfills: the West, Northeast and Southeast New Territories Landfills. However, these are likely to be full by 2019, 2017 and 2015 respectively (Woon and Lo, 2014). At the time of writing, beginning of 2015, the government is proposing to extend the capacity of the three landfills while also building an incinerator on an artificial island near Shek Kwu Cha (TimeOut, 2014a). There is strong opposition to incinerators, due to their environmental implications and a consensus that the government has not tried hard enough to reduce waste at source or encourage recycling.

Only since 2010 have collection points for glass been introduced in Hong Kong. Presently, paper, plastic, metals and glass are recycled. Although recycling rates have risen in recent years, this rate is still much below those of Taiwan and South Korea. In 2012 Hong Kong had recycling rates of only 39 per cent (Cheung, 2014), compared to more than 60 per cent for Taiwan (Chan, 2012). In 2012, only a reported 320,000 tonnes of plastic waste was recycled, compared to 600,000 to 700,000 tonnes of plastic dumped at landfills during the same year (Cheung, 2014). Taiwan and South Korea are two great examples of how implementing effective policies can lead to drastic reductions in domestic waste. Their policies include a levy on rubbish bags, as well as a heightened awareness and consciousness of the general public about recycling. In July 2009 a levy on plastic bags has been introduced among chains or large supermarkets, convenience stores and personal health and beauty product stores, but this has had overall little effect (Lau, 2013).

On the other hand, the problem of construction and demolition waste (C and D waste) from Hong Kong has been solved (from the perspective of Hong Kong) by transporting them to Mainland China for reclamation (Leverett *et al.*, 2007). However, financial and regulatory measures on C and D waste are lacking, and more effort is required to enforce the proper disposal and use of such waste. Leverett *et al.* (2007) suggested that the government should impose stronger regulations on private construction works to reduce or recycle C and D waste.

Figure 5.20 reflects the increase in expenditure for solid waste. This has multiplied by five since 1989, partly as a result of the increasing amount of solid waste disposed, and partly as a result of the increasing costs of disposing such waste, as cheaper alternatives have been exhausted, or face greater popular opposition.

Cost of climate change (−)

Among the environmental items that reflect the output of the economic process, the *Cost of climate change* (Figure 5.21) is by far the largest. In 1968, the per

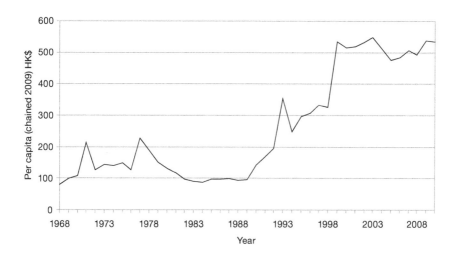

Figure 5.20 Cost of solid waste, Hong Kong

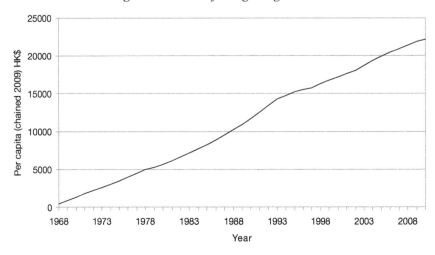

Figure 5.21 Cost of climate change, Hong Kong

capita cost of climate change (based on carbon dioxide (CO_2), methane (CH_4), nitrous dioxide (NO_2), hydrofluorocarbons (HFCs), perfluorocarbons (PFCs), and sulfur hexfluoride (SF_6) emissions) was HK$416. In 2010, it had reached HK$22,193 per person per year, a rate of growth of almost 10 per cent per year. This is an almost identical growth rate as *Personal and public consumption expenditures*, but by far outpacing the rate of growth of GDP. The *Cost of climate change* is very large by any standard, amounting to approximately 14 per cent of *Personal and public consumption expenditure*.

The rate of increase is largely due to the method of calculation: the cost is accumulated, which underlines the fact that this natural capital service cannot be substituted by man-made capital. Hong Kong (as a region of China) is considered a non-Annex I country in the United Nations Climate Change Conference. This means that Hong Kong is not required to meet any specific emission targets. Nevertheless, in 2010 Hong Kong published *Hong Kong's Climate Change Strategy and Action Agenda: Consultation Document*, proposing to reduce carbon intensity by 50–60 per cent by 2020 compared to 2005, as well as other measures related to mitigation and adaptation to climate change (Ng, 2012). However, it is questionable how seriously the government is pursuing its *Climate Change Strategy*. The document was published when the city was moving at full gear with ten major infrastructure projects. In April 2011, when a court overturned the environmental assessment of the three-lane Hong Kong-Zhuhai-Macao Bridge, the Chief Executive claimed that 'a certain political party and politicians make use of legal proceedings or other means, under the excuse of environmental protection or conservatism, to block large-scale projects... they would rather harm Hong Kong's... interests' and warned that more than 70 other projects would be held up because of the ruling (Ng and Cheung 2011, in Ng 2012). Furthermore, one should bear in mind that reducing carbon intensity does not necessarily reduce the total amount of carbon emissions, if the economy grows.

Value of carbon sequestration (+)

The *Value of carbon sequestration* is the only environmental item with a positive sign, but its value remained low throughout the period (Figure 5.22). Hong Kong's country parks were established through the Country Parks Ordinance of 1976. Nowadays, Hong Kong has 24 country parks and one special area, which together cover a total area of 44,300 ha, corresponding to approximately 40 per cent of the total landmass of Hong Kong (AFCD, 2014c). The trees (and soil) in these areas sequester the carbon that is accounted for in this item (the carbon sequestered by water bodies is not included here). The Agriculture, Fisheries and Conservation Department (AFCD) is responsible for the conservation and management of the parks, including tree planting. The AFCD planted 650,000 trees every year between 1991 and 2014, though frequent fires during the dry season destroy some of those trees (AFCD, 2014d).

The total area covered by country parks has been increasing during the period under consideration, and so did the value of the carbon sequestered. In absolute terms, from 1968 to 2010 the value of the carbon sequestered by Hong Kong country parks more than doubled. However, since the population of Hong Kong almost doubled during this period (from 3.8 million in 1968 to 7.1 million in 2010), in per capita terms, the value of the carbon sequestered increased only slightly from HK\$14 in 1968 to HK\$18 in 2010 (in chained 2009 HK\$) (Figure 5.22).

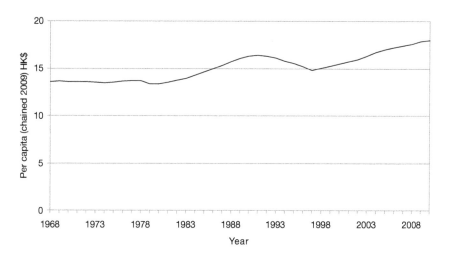

Figure 5.22 Value of carbon sequestration, Hong Kong

Cost of lost wetlands (−)

Records of wetland areas are patchy and incomplete. Since the 1950s (and earlier) many wetlands were dug up for freshwater fish ponds. The area of fish ponds increased from 186 ha in 1954 to up to 2,130 ha in 1986. Such ponds were highly productive for carp and mullet, especially in Yuen Long: until the 1980s, Yuen Long Grey Mullets accounted for 40–50 per cent of the fish produced in local ponds. From the 1970s, fish ponds were gradually sold to developers, who transformed them to other land uses, including housing (Cheung, 2012).

Records state that Hong Kong still had 1,260 ha of wetlands in 1978. The area of wetland dropped to 600 ha in 2000 and to 500 ha in 2003. We use this data, even though we are unable to verify its accuracy. Since 2003, most of the remaining wetland areas are conserved, either de jure or de facto. Because of the distance from the most populated areas of Hong Kong Island and Kowloon Tong, most of the remaining wetlands in Hong Kong are found in the northwestern New Territories. They include streams and rivers, natural marshes, mangroves, intertidal mudflat, as well as artificial fishponds, *gei wais* and reservoirs (AFCD, 2014e), and most notably the 150 ha Mai Po Marshes, listed as a Ramsar site under the Ramsar Convention in 1995, as well as the 60 ha Wetland Park in Tin Shui Wai. The per capita *Cost of lost wetlands* is low throughout the period under consideration, probably because much of the wetland that may originally have existed was lost before 1968 (as one can gauge from Cheung, 2012) (Figure 5.23).

Environmental sub-index

The environmental items (except the *Cost of climate change*) have a lower value than most economic and social items. Rather than being a result of the good

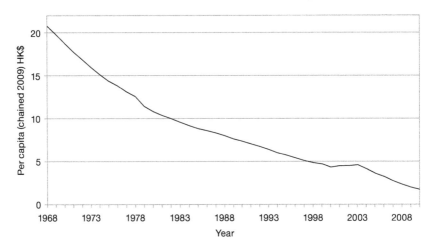

Figure 5.23 Cost of lost wetland, Hong Kong

quality of the Hong Kong environment, the low value is more likely the result of the way in which these items are calculated. The *Costs of air*, *water*, and *noise pollution*, and the *Cost of solid waste*, are all based on government expenditure, and the low value reflect the relatively low emphasis of the Hong Kong government towards environmental conservation, rather than healthy environmental conditions. Similarly, the *Cost of agricultural land degradation* and the *Cost of fishery depletion* are based on the amount of food produced and fish caught, respectively. The little importance of farming, and the fact that overfishing and pollution have reduced the fish catch, contribute to the low costs recorded in the GPI. The *Cost of climate change* is by far the largest environmental cost, accounting for almost 90 per cent of the total costs of the Environmental sub-index since 1986. It is shown on the right axis in Figure 5.24.

The environmental components are made up of 'inputs into the economy' and 'outputs from the economy'. Items included in the input category are the raw materials used in economic activities: non-renewable resources (*Cost of non-renewable resource depletion*), agricultural and fish products (*Cost of agricultural land degradation*, *Cost of fisheries depletion*), and the degradation of wetlands (*Cost of lost wetland*). The items included in the output category are those outputs that result from the production and consumption of products: pollution (*Cost of air pollution*, *Cost of waste-water pollution*, *Cost of noise pollution*), waste (*Cost*

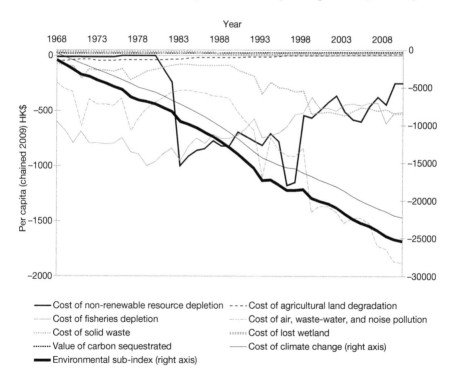

Figure 5.24 Environmental items of the GPI, Hong Kong

of solid waste), and the consequence of greenhouse gasses on the environment (*Cost of climate change* and *Value of carbon sequestered*). It is interesting to note that while the 'input items' have reduced in value, the output items have increased in value. This is likely to be the result of a combination of factors. On the one hand, fewer 'input items' come from Hong Kong, as the salaries and land prices increase, and imports become comparatively cheaper (this would partly explain lower *Costs of agricultural land degradation* and *Cost of fisheries depletion*, but also *Cost of non-renewable resource depletion*). On the other hand, the higher disposable incomes allows for the consumption of more (imported) consumer products, which require additional energy (whose production increase the *Cost of air pollution*), and need to be disposed (increasing the *Cost of solid waste*).

These figures do not include the environmental degradation that consumption by Hong Kong people caused abroad. Most resources (including food, fish, timber products, water, fuel, and minerals) are imported into Hong Kong rather than produced or mined domestically. Likewise, the overwhelming majority of manufactured products consumed in Hong Kong are imported. Finally, some of the waste material (for example construction material and electronic products) are shipped to China, and their detrimental environmental impacts are ignored in Hong Kong. The GPI should ideally include these negative consequences of a country's economic activities on other countries. However, because of lack of data, this never happens. Thus, many 'rich' countries, which import raw materials or manufactured products have a GPI that is higher than it should be. In essence, they export environmental problems. This will be further discussed in Chapter 7.

There is a growing need to promote environmental conservation in Hong Kong. Efforts have been made to restore wetlands and protect mangrove forests. These efforts could be fostered by cooperation with international conservation efforts. For example, it is likely that the ability to register the Mai Po Inner Deep Bay as a Ramsar Site has helped persuading the government to protect the site. International influence plays a role in pushing for improvements. The government can extend this effort to other areas. For example, marine biodiversity has been worsening considerably as a result of over-fishing, pollution and the continued reclamation of fish spawning and nursery grounds (Leverett *et al.*, 2007). Another area that would benefit from additional efforts would be the study and conservation of local flora and fauna species. Its importance is perhaps best highlighted in the Technical Memorandum of the Environmental Impact Assessment Ordinance (EIAO), which was described as 'a major step forward in the formalization of assessment procedures and ecological criteria and [a serious attempt to] improve the professionalism of ecological assessment' by Leverett *et al.* (2007). However, a biodiversity baseline database for Hong Kong is still missing (Leverett *et al.*, 2007). This biodiversity baseline database should provide facts and figures on the state of different biodiversity and ecosystem components, and help better understand how a planned development would affect species and habitats. One of the positive sides, there is the designation of a country park system since the

1970s. However, the original purpose of the system was for water catchment and recreational purposes rather than conservation. As a result, many ecologically rich sites (such as wetlands, and old Feng Shui woodlands) are unprotected (Leverett *et al.*, 2007).

The GPI of Hong Kong: aggregate items

The aggregate results for Hong Kong are shown in Figures 5.25 and 5.26. Figure 5.25 shows the various sub-indices, together with the GPI and the GDP per capita. The results indicate that between 1968 and 2010 Hong Kong's per capita GDP grew by 645 per cent, or an average (compound) growth rate of 4.54 per cent per annum.[2] On the other hand, in the same period, per capita GPI grew by a total of 474 per cent, or an average (compound) growth rate of 3.78 per cent per annum. As expected, overall, the growth rate of per capita GDP exceeded that of per capita GPI, because the social and environmental costs increase more than the economic benefits. This is not inevitable, as the government could make more efforts to improve the environmental and social conditions.

The rate of growth of the GDP stalled from 1997 to 2003 because of the Asian Financial Crisis and SARS, but then continued from 2004 to 2010 (except a drop in 2009). However, the rate of growth of the GPI was slightly different, stalling earlier than the GDP (from 1991 to 1997) and then again after the Asian Financial Crisis (from 2002 to 2005), and finally dropping from 2008 to 2010. Overall, the gap between the GDP and the GPI continued to increase. Only during a brief

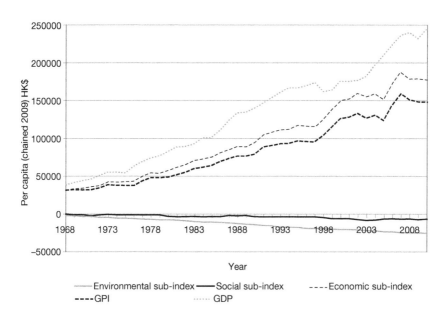

Figure 5.25 Comparison of the different sub-indices, the GPI and the GDP, Hong Kong

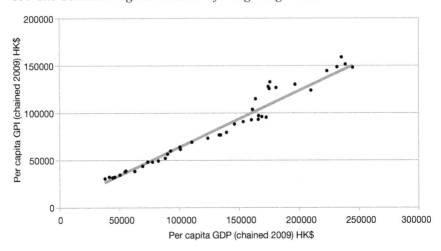

Figure 5.26 Per capita GDP versus per capita GPI: Hong Kong, 1968–2010

period, following the Asian Financial Crisis, did the gap between the GDP and the GPI drop. It is quite possible that if the GDP stopped growing, the GPI would continue to grow, because some environmental and social items would improve. This will be further discussed in Chapter 8.

The results are overall rather unexpected, as the GPI of other countries stopped growing in the 1970s or 1980s, while the GDP kept increasing (see Chapter 7). This may suggest that Hong Kong's economic development may have grown at a rate that is still relatively sustainable and that it is able to improve the welfare of its people. However, in Chapter 7 we will also discuss alternative arguments, in particular the impact of 'offshoring' particular sectors of the Hong Kong economy, including the pollution generated, and the raw materials needed, to China. Indeed, the increase in the GPI after 1997 can be related to the shift of the Hong Kong economy, and the export of companies to the Guangdong province (see Chapter 7).

Figure 5.26 provides a different representation of the same trend, with a close to linear relationship between the GDP and the GPI, but a much faster growth of the GDP, compared to the GPI.

Conclusions

In this chapter we have reported the values of the various items included in the GPI. Some of these values are very small, and are inconsequential in the aggregated GPI values. This includes, for example, *Cost of family breakdown*, *Cost of agricultural land degradation*, and *Cost of carbon sequestration*. In other countries, these may be important, and for the sake of consistency we have chosen to retain them in our study. Overall, the GPI has increased throughout the study period, because the growing economic items have far outweighed the dropping environmental and social items. This is a surprising trend. The GPI of most nation states has been either steady or dropping since the late 1970s or early 1980s (Chapter 7). In Chapter 7 we will discuss why the GPI of Hong Kong has kept increasing. In particular, Hong Kong's GPI figures do not account for the environmental costs of production of (imported) goods. However, overall Hong Kong seems to have achieved some remarkable successes in spite of sometimes difficult conditions, partly due to its closeness to the People's Republic of China and a preferential treatment by its government. In the next chapter we present the GPI of Singapore. In Chapter 7 we compare the two.

Note

1   Marine vessels are the largest contributor to Hong Kong air woes, rather than idling engines or power plants. Cargo ships use low-quality, high-sulfur fuel when navigating on the high sea, with a sulphur content that is up to 2,000 times higher than that used in motor vehicles. These ships are already carrying low-sulphur fuel, which they must use when they enter the territorial waters (200 nautical miles from the shore) of most countries. Yet, Hong Kong has no such requirement, so these tankers can use high-sulphur fuel in the Hong Kong port. This may go a long way to explain why Hong Kong has a standard for Sulphur dioxide that is 625 per cent higher than that of the WHO. In 2011, ships accounted for 54 per cent of total sulphur dioxide emissions in Hong Kong. Hong Kong introduced legislation that requires vessels to use extra-low-sulphur diesel upon arriving at berths at Kwai Chung Container Terminal from January 2015 (SCMP, 2013b), but not within 200 nautical miles from the shore.
2   All figures in this discussion are in chained 2009 HK$.

References

AFCD. (2014a). Agriculture in HK. Agriculture, fisheries and conservation department, retrieved 27 November 2014 from www.afcd.gov.hk/english/agriculture/agr_hk/agr_hk.html

AFCD. (2014b). Fisheries. Agriculture, fisheries and conservation department, retrieved 27 November 2014 from www.afcd.gov.hk/misc/download/annualreport2013/en/fisheries.html

AFCD. (2014c). Country Parks and Conservation. Agriculture, fisheries and conservation department, retrieved 27 November 2014 from www.afcd.gov.hk/english/country/cou_lea/the_facts.html

AFCD. (2014d). Useful Statistics. Agriculture, fisheries and conservation department, retrieved 27 November 2014 from www.afcd.gov.hk/english/country/cou_lea/cou_lea_use/cou_lea_use.html

AFCD. (2014e). General Information about Wetland. Agriculture, fisheries and conservation department, retrieved 27 November 2014 from www.afcd.gov.hk/english/conservation/con_wet/con_wet_abt/con_wet_abt_gen/con_wet_abt_gen_wet.html

Badalamenti, F., Pipitone, C., D'Anna, G. and Fiorentino, F. (2012). The trawling ban in Hong Kong's inshore waters – A round of applause and a plea to learn from others' mistakes. *Marine Pollution Bulletin, 64*(8), 1513–1514.

Bouffard, D., Cook, S., Eisenberg, S. and Mowris, R. (2013). Investigating the Relationships between Urban Design, Microeconomics, and Livability: A Case Study of Hong Kong.

Broadhurst, R., Bouhours, B., Bacon-Shone, J., Zhong, L. Y. and King Wa, L. (2010). Hong Kong United Nations International Crime Victim Survey: Final Report of the

2006 Hong Kong UNICVS. Hong Kong: The University of Hong Kong, Social Sciences Research Centre and Canberra: The Australian National University, Centre of Excellence in Policing and Security.

Carney, J. (2011). HK men twice as likely to overwork. *South China Morning Post*. 10 November 2011. Accessed online: www.scmp.com/article/984381/hk-men-twice-likely-overwork on 14 November 2014.

CEDD. (2014). Economic Geology – Minerals and Mining in Hong Kong. Hong Kong Geology – A 400-Million Year Journey. Hong Kong: Civil Engineering and Development Department (CEDD). Accessed 14 October 2014 from http://hkss.cedd.gov.hk/hkss/eng/education/GS/eng/hkg/indexe.htm

Chan, A. C., Cheung, S. Y. and Lai, T. C. (2014). Widening of a poverty gap: A condition of governance crisis in Hong Kong. *Advances in Applied Sociology*, 2014.

Chan, R. (2012). EPA touts Taiwan's successful recycling policy. Taiwan Today. Accessed 14 November 2014 on http://taiwantoday.tw/ct.asp?xItem=195110&ctNode=413

Cheung, C-F (2014). Estimates of the amount of Hong Kong rubbish being recycled are plain rubbish. *South China Morning Post*. Accessed online on 14 November 2014 at www.scmp.com/news/hong-kong/article/1415979/recycling-figures-plain-rubbish

Cheung, S. C. H. (2012). History of Freshwater Fish Farming the New Territories, Hong Kong. Presentation at the Chinese University of Hong Kong.

Craig, R. S. (2011). *Determinants of property prices in Hong Kong SAR: implications for policy*. International Monetary Fund.

DeGolyer, M. E. (2008). *Hong Kong Silent Epidemic*. Hong Kong: Civic Exchange.

Delang, C. O. and Cheng, W. T. (2013). Hong Kong people's attitudes towards electric cars. *International Journal of Electric and Hybrid Vehicles*, 5(1), 15–27.

EPD. (2012). River Water Quality in Hong Kong in 2012. Hong Kong: Environmental Protection Department.

EPD. (2014). Hong Kong's Air Quality Objectives. Hong Kong: Environmental Protection Department. Accessed online on 14 November 2014 at www.epd.gov.hk/epd/english/environmentinhk/air/air_quality_objectives/air_quality_objectives.html

Gottret, P., Gupta, V., Sparkes, S., Tandon, A., Moran, V. and Berman, P. (2009). Protecting pro-poor health services during financial crises: Lessons from experience. *Advances in Health Economics and Health Services Research*, 21, 23–53.

HKIE. (2009) Blueprint for a Better Future: An engineering perspective on Hong Kong's infrastructure. Hong Kong: The Hong Kong Institution of Engineers.

Ho, P. S. Y. (2007). Eternal mothers or flexible housewives? Middle-aged Chinese married women in Hong Kong. *Sex roles*, 57, 249–265.

Hong Kong Census and Statistics Department. (2012). *Annual Digest of Statistics*. Hong Kong: Hong Kong Census and Statistics Department.

Hong Kong Journal. (2006). *An Interview with Donald Tsang, Hong Kong's Chief Executive*, Accessed 25 November, 2014, at www.hkjournal.org/archive/2006_summer/tsang.html

Hung, S. L., Kung, W. W. and Chan, C. L. (2004). Women coping with divorce in the unique sociocultural context of Hong Kong. *Journal of Family Social Work*, 7(3), 1–22.

Lai, L. W. C., Chau, K. W., Ho, D. C. and Lin, V. Y. (2006a). Impact of political incidents, financial crises, and severe acute respiratory syndrome on Hong Kong regulators and developers. *Environment and Planning B: Planning and Design*, 33(4), 503.

Lai, L. W. C., Chau, K. W., Ho, D. C. and Lin, V. Y. (2006b). Impact of political incidents, financial crises, and severe acute respiratory syndrome on Hong Kong property buyers. *Environment and Planning B: Planning and Design*, 33(3), 413.

Lau, E. (2013). Extension of plastic bag levy can educate people about cutting waste. *South China Morning Post*. Accessed 14 November from www.scmp.com/comment/insight-opinion/article/1336566/extension-plastic-bag-levy-can-educate-people-about-cutting

Lee, S., Guo, W. J., Tsang, A., Mak, A. D., Wu, J., Ng, K. L. and Kwok, K. (2010). Evidence for the 2008 economic crisis exacerbating depression in Hong Kong. *Journal of affective disorders*, *126*(1), 125–133.

Leung, L. C. and Chan, K. W. (2012). Understanding the masculinity crisis: implications for men's services in Hong Kong. *British Journal of Social Work*, bcs122.

Leverett, B., Hopkinson, L., Loh, C. and Trumbull, K. (2007). Idling Engine: Hong Kong's environmental policy in a ten year stall, 1997–2007. Civic Exchange: Hong Kong.

Liao, H. F. and Chan, R. C. (2011). Industrial relocation of Hong Kong manufacturing firms: towards an expanding industrial space beyond the Pearl river delta. *GeoJournal*, *76*(6), 623–639.

Lui, H. K. (2013). *Widening income distribution in post-handover Hong Kong*. London: Routledge.

Meghir, C. and Pistaferri, L. (2011). Earnings, consumption and life cycle choices. *Handbook of Labor Economics*, *4*, 773–854.

Ng, K. C. and Cheung, G. (2011). Certain party is hurting city: Tsang. *South China Morning Post* (20 May 2011).

Ng, M. K. (2012). A critical review of Hong Kong's proposed climate change strategy and action agenda. *Cities*, *29*(2), 88–98.

Planning Department. (2013). Land Utilization in Hong Kong. In www.pland.gov.hk/pland_en/info_serv/statistic/landu.html

SCMP. (2014). Foreign domestic workers in Hong Kong. *South China Morning Post*. Accessed on 14 September 2014 from www.scmp.com/topics/foreign-domestic-workers-hong-kong

Siu, A. and Wong, Y. R. (2004). Economic Impact of SARS: The Case of Hong Kong*. *Asian Economic Papers*, *3*(1), 62–83.

Song, H. and Lin, S. (2009). Impacts of the financial and economic crisis on tourism in Asia. *Journal of Travel Research*.

Sullivan, P. L. (2005). Culture, divorce, and family mediation in Hong Kong. *Family Court Review*, *43*(1), 109–123.

Thach, T. Q., Wong, C. M., Chan, K. P., Chau, Y. K., Chung, Y. N., Ou, C. Q. and Hedley, A. J. (2010). Daily visibility and mortality: assessment of health benefits from improved visibility in Hong Kong. *Environmental Research*, *110*(6), 617–623.

Thomas, D., Beegle, K., Frankenberg, E., Sikoki, B., Strauss, J. and Teruel, G. (2004). Education in a Crisis. *Journal of Development Economics*, *74*(1), 53–85.

TimeOut. (2014a). Dissecting the Shek Kwu Chau incinerator debate. TimeOut Hong Kong, 9 July 2014. Accessed online on 14 November 2014 from www.timeout.com.hk/big-smog/features/67917/dissecting-the-shek-kwu-chau-incinerator-debate.html

TimeOut. (2014b). Hong Kong's new Air Quality Health Index. TimeOut Hong Kong, 8 January 2014. Accessed 14 November 2014 at www.timeout.com.hk/big-smog/features/63300/hong-kongs-new-air-quality-health-index.html

Wang, E. C. (2002). Public infrastructure and economic growth: a new approach applied to East Asian economies. *Journal of Policy Modeling*, *24*(5), 411–435.

Wong, D. (1996). Housewife role and women's psychological well-being: The case of Tuen Mun. *Hong Kong Journal of Social Work*, *30*(1), 48–56.

Woon, K. S. and Lo, I. (2014). Analyzing environmental hotspots of proposed landfill extension and advanced incineration facility in Hong Kong using life cycle assessment. *Journal of Cleaner Production, 75*, 64–74.

World Bank. (2014). GINI index. Accessed 19 October 2014 at http://data.worldbank.org/indicator/SI.POV.GINI

WSJ. (2012). Businesses' Biggest Hong Kong Complaints: Pollution, Schools. *The Wall Street Journal*. 8 May 2012.

YLDC. (2011). *2011 Population Census Fact Sheet. Yuen Long District Council.* In Hong Kong Census and Statistics Department.

6 The Genuine Progress Indicator of Singapore: results

With: Enoch Yin Lok Lee (Department of Geography, Hong Kong Baptist University)

Introduction

Singapore was once a British naval military base and a commercial trading outpost (Cahyadi *et al.*, 2004). Today, Singapore is a large exporter of manufactured goods, a world port and an international finance centre (Lim and Pang, 1986), and is known as one of four 'Asian Tigers'. From 1968 to 2010 Singapore experienced considerable growth. Singapore gained independence from Malaysia in 1965, three years before our analysis starts. When it gained independence, Singapore faced a myriad of uncertainties. Within the region, the Konfrontasi was ongoing and the conservative UMNO faction strongly opposed the separation, while Singapore faced the dangers of attack by the Indonesian military (Poulgrain, 1998). Domestically, the pressing problems were unemployment (with an unemployment rate of about 10 per cent), lack of adequate housing, a low education standard, and the lack of natural resources to exploit to spur economic growth. At the same time, Singapore had lost access to Malaysia's raw materials and domestic market (Krause, 1987; Turnbull, 2009). Independence also meant Singapore had to change its development strategy. While initially Singapore attempted to use an import substitution industrialization (ISI) strategy – aiming to increase self-sufficiency by protecting local industries from foreign competition – it soon focused on producing labour-intensive manufacturing products for export (Lim and Pang, 1986). To do so, Singapore made itself more attractive to multinational corporations and international investors. The strategy proved very successful, and by 1978 the economy was growing rapidly and the unemployment rate had declined to 3.59 per cent.

Due to such rapid growth, the labour market tightened, and pressure to raise wages intensified (Turnbull, 2009). Singapore solved this problem by shifting away from labour-intensive manufacturing industries towards higher value-added, knowledge-intensive service industries, such as research and development, engineering design and computer software services. In 1981, the Singapore government formed the National Computer Board (NCB) to develop a workforce with adequate information and computer technology (ICT) skills. It also attempted to draw in more foreign IT corporations (Cahyadi *et al.*, 2004).

Throughout the 1990s, Singapore continued restructuring, investing S$2 billion from 1991 to 1995 and S$4 billion from 1996 to 2000 into improving technology (Wong, 2002). The service and manufacturing industries were identified as the twin pillars of the economy, and many investments were extended to the services sector (Cahyadi *et al.*, 2004). The Economic Development Board (EDB) further strengthened its focus on the chemical, electronics and engineering industries, while developing its pharmaceutical, biotechnology and medical technology sectors. In 1994, the Singapore-Johor-Riau (SIJORI) growth triangle was established to make the region even more attractive to foreign investors (Cahyadi *et al.*, 2004).

During the last 15 years this study is concerned with, the Singapore economy, as measured by the GDP, has experienced somewhat of a rollercoaster, with high growth of around 7–8 per cent during some years (1995–1997, 2000, 2004–2007), followed by a shrinking economy during other years (1998, 2001, 2009). The economic instability was partly due to regional crises, such as SARS and the Asian Financial Crisis, economic downturns in other major developed economies, and a global slump in the demand for electronics products, which reduced demand for Singapore's electronics manufacturing industries (Das, 2010). Singapore was also affected by the outbreak of the Severe Acute Respiratory Syndrome (SARS) in 2003, and consequently faced heavy expenses in prevention and relief (Rossi and Walker, 2005). SARS also had a major negative impact on Singapore's tourism industry, further weakening the economy (Rossi and Walker, 2005).

As the economy recovered, Singapore increased expenditure on education and in particular investment in knowledge- and innovation-intensive industries, a sector in which the government decided to invest more than S$13 billion in research and development in 2006. The National Research Foundation (NRF) was founded that year to coordinate and implement nationwide research and innovation strategies (Yeoh, 2006; Fuyuno and Cyranoski, 2006; Shatkin, 2014).

Singapore's economic development has been extraordinary, having demonstrated substantial growth over the past few decades. But did welfare keep pace with the rate of economic growth? In the following pages we look at the GPI of Singapore, from 1968 to 2010. We start with discussing the individual sub-indices, starting with the economic, and then turning to the environmental and the social. Assessing the individual items of the GPI will help us better understand Singapore's development trend during the last decades. We then present the aggregate GPI, and compare it to the GDP figures.

Economic items

Personal and public consumption expenditure (+)

Personal and public consumption expenditure forms the bulk of the GPI, as it forms the bulk of the GDP. Figure 6.1 shows the changes over the four decades under consideration. As can be seen from the figure, there has been a constant, regular increase, from S$5,494 in 1968 to S$28,325 in 2010, or 4 per cent a year. The Asian Financial Crisis has had only a very limited impact to Singapore, especially compared to Hong Kong, because Singapore used the exchange rate and wage instruments effectively during the crisis (Ngiam, 2000). The HK$ is pegged to the US$ and during the Asian Financial Crisis the Hong Kong government decided to retain the peg unchanged rather than depreciating the currency. On the other hand, Singapore decided to quickly depreciate the S$ in response to the loss of export competitiveness arising from the collapse of the regional currencies. As the crisis dragged on, Singapore decided not to tinker with the nominal exchange rate but instead worked towards direct cost-cutting measures to maintain its competitiveness, such as wage and operating cost reductions (Ngiam, 2000).

Defensive and rehabilitative expenditure (−)

Figure 6.2 reports the *Defensive and rehabilitative expenditure* from 1968 to 2010. As for the *Personal and public consumption expenditure* there has been a regular increase, from S$1,321 in 1968 to S$6,063 in 2010. The rate of growth is slightly below that of *Personal and public consumption expenditure*, at 3.7 per cent a year. There is a considerable difference between Hong Kong and Singapore, Hong Kong starting with a much lower expenditure than Singapore in 1968, but ending with a 45 per cent higher expenditure in 2010.

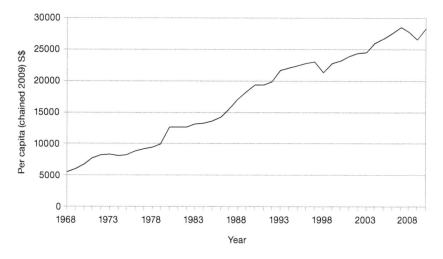

Figure 6.1 Personal and public consumption expenditure, Singapore

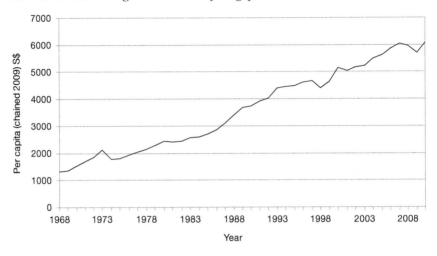

Figure 6.2 Defensive and rehabilitative expenditure, Singapore

Expenditure on consumer durables, and Services from consumer durables (− and +)

Expenditure on consumer durables increased from 1968 to 1994, and since then has remained rather stable, or dropped (Figure 6.3). On the other hand, *Services from consumer durables* has increased until 2003, before dropping. This is due to the ways in which the *Services from consumer durables* is calculated. Once the expenditure is made, the benefits from consumer durables are expected to last for seven years, and are spread over that period, so any drop in expenditure is felt gradually. *Expenditure on consumer durables* is subtracted from the GPI, while *Services* are added. The shape of these curves is similar to that of Hong Kong, even though per capita expenditure in Hong Kong is higher.

Weighted adjusted consumption expenditure

Figure 6.4 presents graphically the *Weighted adjusted consumption expenditure* with all its components, except the *Income distribution index*, which is used to weight it. *Weighted adjusted consumption* experienced fast and steady growth until the late 1990s. However, from 1999 it hardly changed. What may have prompted this slowdown is the Asian Financial Crisis of 1997, and the financial crisis of 2008. While Hong Kong has been able to benefit from the growth in neighbouring China, and experience considerable growth in *Weighted adjusted consumption expenditure* from 2003 onward, Singapore's *Weighted adjusted consumption expenditure* seems to have stabilized at around S$20,000 per capita since 1999.

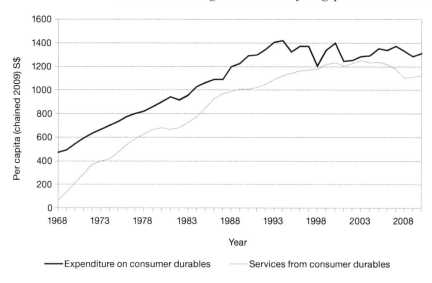

Figure 6.3 Expenditure on consumer durables and the Services from consumer durables, Singapore

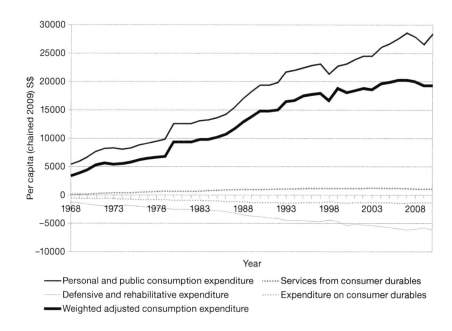

Figure 6.4 Weighted adjusted consumption expenditure and its components, Singapore

Income inequality, which is used to weigh *Weighted adjusted consumption expenditure*, has increased during the study period. Information on inequality is available only for scattered years. The GINI coefficient was initially relatively low, at 0.418 in 1982–83, 0.402 in 1988–89 (Fields, 1994), and 0.425 in 1998 (Watkins, 2007), although it recently increased quite considerably, to 0.472 in 2010 (DOS, 2012). This figure only represents income from work (with employers' CPF contributions) and is slightly misleading because it does not take into account government benefits or taxes. Furthermore, slightly more than 80 per cent of Singaporeans live in subsidised Housing Development Board (HDB) apartments (Ong and Tim, 2014), which reduces inequality – though not to the extent that public housing reduces inequality in Hong Kong, because in Singapore more people are able to benefit from public housing, and because the rent is considerably higher than it is in Hong Kong. After adjusting for government transfers and taxes, the Gini coefficient in Singapore in 2010 fell from 0.472 to 0.451 (DOS, 2012). In spite of this drop, Singapore has one of the highest levels of inequality in the developed world. Donaldson *et al.* (2013: 60) point out that 'inequality in Singapore is compounded in comparison to countries with similar per capita income by the low and falling real wages of the bottom 20 per cent of employed residents relative to that of other wage earners. The period 1998 to 2010 saw the real median incomes of employed residents in this quintile fall by approximately 8 per cent, while incomes of those in the top 20 per cent increased by 27 per cent'.

Services yielded from fixed capital (+)

Services provided by fixed capital have steadily increased from 1968 to 2003, reflecting the investments that the Singapore government has been making in infrastructure (Figure 6.5). In 1968 the figure for *Services provided by fixed capital* was of S$475 per person. By 2003 this had increased to S$12,589. From 2004 it started to drop, and by 2010 it was S$10,691 per capita, a drop of 15 per cent. This item includes infrastructural capital such as roads, highways, bridges, schools and hospitals, and is estimated per capita. The drop in per capita values is due to the fact that the population of Singapore grew rapidly, from 4.115 million in 2003 to 5,077 million in 2010, a growth of 23 per cent in only seven years. In absolute terms *Services provided by fixed capital* remained stable after 2003.

The population of Singapore has increased at an average rate of almost 2.2 per cent per year from 1968 to 2002, and of more than 3 per cent per year from 2003 to 2010, as the government encouraged immigration. This can be compared to an average growth rate of less than 1.5 per cent per year for Hong Kong, from 1968 to 2010.

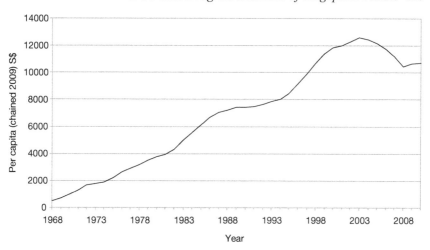

Figure 6.5 Services yielded from fixed capital, Singapore

Change in net foreign assets (+/−)

Change in net foreign assets does not show a particular trend. The first 16 years (from 1968 to 1984) experienced low investments. Singapore became an independent country only in 1965, and the low initial values, S$241 per capita in 1968 may be related to the fact that it had only recently gained independence. This period was followed by rapid growth (from 1985 to 1993), a period of decline and lower investment (a ten-year period from 1994 to 2003) followed again by rapid growth (from 2004 to 2007). *Change in net foreign assets* reflects investments by foreign companies in Singapore, as well as investments of Singaporean companies abroad. Singapore investments abroad include those of government agencies, such as Temasek, an investment company owned by the Government of Singapore (Bernstein *et al.*, 2013), with a net portfolio value at S$223 billion in 2014, focused 69 per cent outside Singapore, and for 41 per cent in Asia ex-Singapore (Temasek, 2014). Interpretations of the data are difficult because both investments into and out of Singapore are included in the figure. The lack of trend of the curve reflects the two trends of inward investment by foreign investors, and outward investment by Singaporeans, mainly in Southeast Asia.

The Singapore government has a long history of efforts to attract foreign investment. The importance of this foreign investment to Singapore's economic development must not be understated. Lacking natural resources, Singapore has relied heavily on its vantage point as an important trading hub, and thanks to its favourable geographic location and its political stability, it has been able to attract the headquarters of multinational corporations (MNCs) which manufacture in the region (Yeung, 2001; Thite *et al.*, 2012; Lim, 2010). Gradually, it has attracted also high-tech research and development sectors with

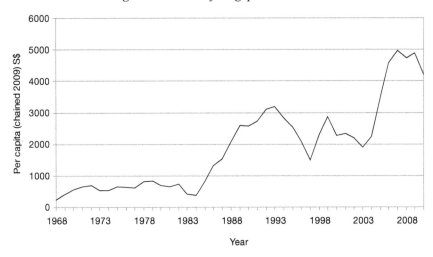

Figure 6.6 Change in net foreign assets, Singapore

high value-added manufacturing. The investment in Singapore has been encouraged by favourable capital export laws, compared to the neighbouring countries, and the encouragement of foreign talents. At the same time, investments in its own fixed assets, in the form of infrastructure, supported not only the local economy but also encouraged further foreign direct investment (Peebles and Wilson, 1996; Li, 2002). Its emphasis on education and research and development can also be said to have the same goal of providing highly qualified personnel to foreign investors.

Economic sub-index

Figure 6.7 shows all the economic items together (except the aggregated *Weighted adjusted consumption expenditure*). The items that are added to the GPI are plotted above the x axis; those that are subtracted from the GPI (such as *Defensive and rehabilitative expenditure*) are plotted below the x axis. Unlike in Hong Kong, there has not been a drop in *Personal and public consumption expenditure* due to the Asian Financial Crisis and SARS, although the rate of growth has slowed down from the early 1990s. These results will be complemented to those of the social and environmental indices, presented in the next pages, to assess the changes in the levels of welfare in Singapore.

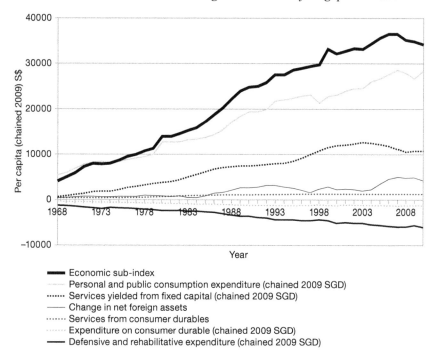

- Economic sub-index
- Personal and public consumption expenditure (chained 2009 SGD)
- Services yielded from fixed capital (chained 2009 SGD)
- Change in net foreign assets
- Services from consumer durables
- Expenditure on consumer durable (chained 2009 SGD)
- Defensive and rehabilitative expenditure (chained 2009 SGD)

Figure 6.7 Economic items of the GPI, Singapore

Social items

Value of non-paid household labour (+)

The *Value of non-paid household labour* was very high at the beginning of the period under consideration, but dropped rapidly, and by 1973 it was only S$379 per person (Figure 6.8). For the remainder of the period it lingered between S$158 and S$385. This drop reflects the decreasing number of people in Singapore that do not engage in paid employment. After gaining independence in 1965, the labour-intensive manufacturing sector grew, and formerly inactive members of the workforce were encouraged to work (Lim and Pang, 1986). Thus, during industries in the late 1960s and early 1970s the decline of non-paid household labour was due largely to the significant increase in the numbers of housewives joining the labour-intensive manufacturing sector. Subsequent changes in social attitudes and the rising costs of Singapore's urban environment also led to greater female participation in the labour force (ibid.).

As in the case of Hong Kong, much household work is being performed by foreign domestic workers, mostly from the Philippines and Indonesia. At the end of 2010 there were 201,000 foreign domestic workers in Singapore, compared to 140,000 in 2002 (MoM, 2014). This means that about 20 per cent of households

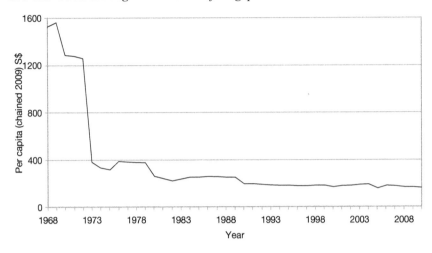

Figure 6.8 Non-paid household labour, Singapore

employed a foreign domestic helper in 2010. Work conditions are usually worse than in Hong Kong, with a lower wage and less time off. Only since January 2013 foreign domestic workers have had the right to one day off a week. The employer and the domestic worker can mutually agree on compensation in-lieu of rest days, and a survey in 2013 found that only about a third of foreign domestic workers declared to have at least one day off a week (Tan, 2013)

Value of volunteer labour (+)

From 1993, the *Value of volunteer labour* overtook that of non-paid household labour (Figure 6.9). The substantial growth in the *Value of volunteer labour* was mainly due to the fact that we use the average monthly wages in the community sector to estimate their economic value. As average monthly wages increased from $791 in 1981 to $4,292 in 2010, the *Value of volunteer labour* increased proportionally. Its increase may also be due to a greater recognition of the psychological benefits gained from volunteer work (as previously mentioned in the study of Hong Kong). These benefits take the form of active aid to those in need, or a pathway towards active engagement in society (Wong and Foo, 2011). We should mention that *Value of volunteer labour* is estimated with the official data of Singapore's Yearbooks of Statistics, and therefore our data does not take into account unregistered volunteers. Hence, the increasing value may also be due to a gradually more thorough recording of volunteer labour during the time under consideration.

Cost of security and external relations (−)

The most important social item is the *Cost of security and external relations* (Figure 6.10). For Hong Kong, only the cost of crime is included because Hong

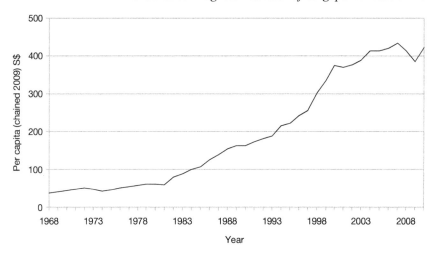

Figure 6.9 Value of volunteer labour, Singapore

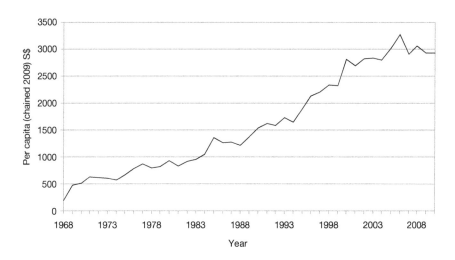

Figure 6.10 Cost of security and external, Singapore

Kong does not have a military force of its own. On the other hand, Singapore has one of the highest percentages of military expenditure to GDP, with the item accounting for 4 to 5 per cent of GDP in the last decades (SIPRI, 2014). This large expenditure may be justified by the fact that Singapore is surrounded by two very large countries, with which it sometimes has tense relations (Tan, 1999; Ganesan, 2005). However, one can hardly argue that military expenditure increases welfare, so this item is subtracted from the GPI.

Cost of security and external relations includes the categories 'Operating expenditure on defense, justice and police' and 'Development expenditure on defense, justice and police', until these were renamed 'security and external relations' in the 1995 Yearbooks of Statistics. This item steadily increased from 1968 to 2000, reaching close to S$3,000 per person per year in 2000, when it somehow stabilised on a per capita basis (although high population growth means that it kept increasing in absolute terms).

Cost of unemployment and underemployment (−)

The *Cost of unemployment and underemployment* was low for most years from 1968 to 1997, before rising sharply from 1998 to 2000 during the Asian Economic Crisis (Figure 6.11). Since 2000 it has been relatively constant at about S$1,000 per person per year.

Until the late 1990s, Singapore has had persistently low unemployment rates of 1.6–2.6 per cent, and severe labour shortages in all sectors and at all skill levels. Those who wanted to work had no difficulty finding employment because the number of jobs exceeds the supply of workers (Van Dyne and Ang, 1998). Kee and Hoon (2005) argue that overall the massive capital accumulation in the export sector helped maintaining a low rate of unemployment in Singapore since the late 1960s. In particular, Singapore has been successful in attracting new types of foreign direct investments in manufacturing, thereby developing and modernizing its economy. In the 1960s and early 1970s, most foreign direct investments were in garments and textiles, as well as in simple assembly-line electronics. During this period, Multinational Corporations (MNCs) brought in parts from other countries to assemble and re-export. From the mid-1970s, these labour-intensive activities were relocated to other countries in Southeast

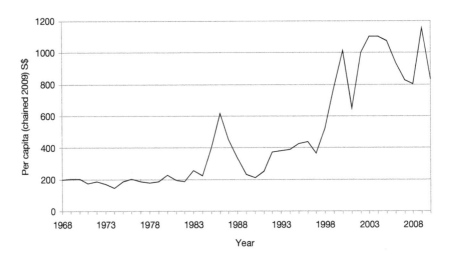

Figure 6.11 Cost of unemployment, Singapore

Asia, and foreign direct investments moved higher up the value chain, focusing on such things as disk drivers and semiconductors, as well as trade- and management-related sectors (Kee and Hoon, 2005). As the economy was modernized, the government put much investment into educating the workforce and attracting foreign talent to Singapore. This way, Singapore could avoid competing with the neighbouring countries for the more labour-intensive industries, thereby fostering employment and economic growth (Chiu *et al.*, 1997; Keng *et al.*, 2004).

In the 2000s the situation worsened somehow, with an official unemployment rate of 4 per cent in 2003 (Rajan, 2009). The official unemployment rate is kept low by the fact that Singapore does not have unemployment benefits and the limited financial help that people may receive (up to S$530 a month for a family of two adults and two children in 2000) is 'administered strictly to exclude applicants who may receive support from their family, friends and relatives' (Wai-lam, 2000: 6).

Cost of overwork (−)

The second most important item is the *Cost of overwork* (Figure 6.12). As in the case of Hong Kong, overwork is very common in Singapore, and its cost has been increasing steadily from 1968 to 2000, when it stabilized. As in the case of Hong Kong, the roots of the overwork problem in Singapore may be found in its elite-oriented business culture in a competitive environment. In such an environment, company employees usually have to work long hours to show commitment to the company. Work pressure is compounded when conflicts arise between demands from work and family life. Chan *et al.* (2000) argued that Singaporeans have been educated and are encouraged to commit themselves to both work and family. However, once internalized 'the sense of commitment to both social roles may

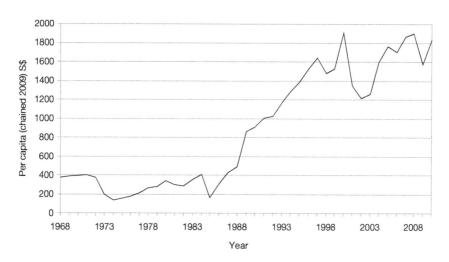

Figure 6.12 Cost of overwork, Singapore

generate a psychological dilemma when the overall demands are multiple and incompatible' (Chan *et al.*, 2000: 1430). The economic problems that accompanied the Asian Economic Crisis and SARS are also reflected in this item, since the costs of overwork have dropped for a brief period, from 2001 to 2003. Since 1997, they have stabilized at around S$1,600 per person per year.

Cost of family breakdown (−)

Cost of family breakdown is still very small in Singapore, reflecting the importance of the family unit in Singapore society, and the pressures there are against divorce (Figure 6.13). The low divorce rate in Singapore, of 7.6 per 1,000 married resident males and 7.3 per 1,000 married resident females, could also be due to campaigns to promote family values from 1994 (Keng *et al.*, 2004). Keng *et al.* (2004) found that family life is one of the domains that Singaporeans are more satisfied with, and Singapore culture is said to be very family-oriented.

Social sub-index

The social conditions of Singapore have improved greatly since the end of British colonialism, and studies have illustrated this improvement in the form of longer life expectancy, the elimination of slums and the decrease in poverty, improvements in education and the living environment, as well as better interracial harmony (Lim, 2010). Kuan *et al.* (2010) shows that Singaporeans in general are quite satisfied with life, as indicated by the high ratings they have given in satisfaction surveys to an average of 12.35 out of 16 life domains.[1] The top five domains are marriage, family life, friendships, public safety, and housing; while the least satisfactory are household income and the social welfare system. Lower-income

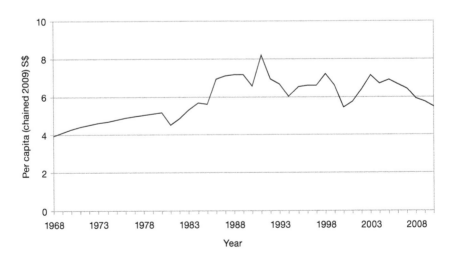

Figure 6.13 Cost of family breakdown, Singapore

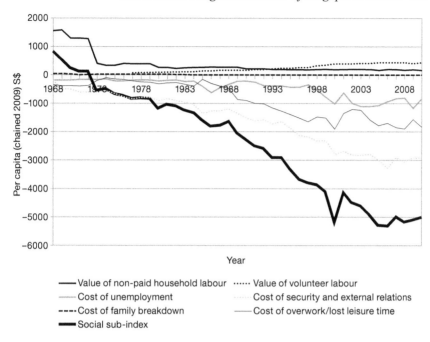

Per capita (chained 2009) S$

Year

——— Value of non-paid household labour ⋯⋯ Value of volunteer labour
——— Cost of unemployment ⋯⋯ Cost of security and external relations
- - - - Cost of family breakdown ——— Cost of overwork/lost leisure time
▬▬ Social sub-index

Figure 6.14 Social items of the GPI, Singapore

groups in particular appear to be unsatisfied with household incomes, jobs, education, social welfare, and the democratic system (Kuan *et al.*, 2010). However, the GPI reveals other facets of the social life in Singapore (Figure 6.14). In particular, the GPI points to considerable costs that accompany economic development and do not increase the well-being of Singapore people. As can be seen, the social items with a negative sign are growing much more rapidly than the few social items with a positive sign. Overall, the situation has been worsening considerably during the period under consideration.

Environmental items

We now turn to the environmental items of the GPI of Singapore. As with the economic and social items, we first present each item individually, and then the aggregate data. Most environmental items have a negative sign, because economic activities involve the transformation, consumption and disposal of natural resources. In the following pages, it will become apparent that similar to the social costs, most environmental costs have also been increasing during the period under consideration, as the economy has expanded.

Cost of non-renewable resource depletion (−)

We start with the *Cost of non-renewable resource depletion* (Figure 6.15). The GPI only includes the cost of minerals mined in the national territory. Imported minerals are not included. This cost has been increasing rather steadily from the 1960s, first slowly, and then more rapidly, although there was a slump in the 1990s. The overall *Cost of non-renewable resource depletion* is relatively low. However, in the mid-2000s, the costs jumped to become the third largest contributor to the environmental sub-index.

In Singapore there was little mining throughout the period under consideration. However, granite quarrying was once a major industry on Pulau Ubin (Ubin island), where the first quarry was established in the 1800s. Ubin's granite was used to make concrete (among others, it was used to build the Horsburgh Lighthouse and Raffles Lighthouse, at the two extreme points of Singapore) and for reclamation. In the 1990s, the Aik Hwa Granite Quarry was still supplying 30 to 40 per cent of Singapore's requirements. The granite industry supported a relatively large population on Ubin: at one time, Aik Hwa Granite Quarry employed about 100 workers, most of who lived on the island, and indirectly supported a wide range of supporting cottage industries. Quarrying gradually ceased when it became profitable to mine granite below sea level. Aik Hwa was one of the last quarries to close, in 1999 (Wild Singapore, 2014). In 2007 there were calls to restart the granite quarry on Pulau Ubin after Indonesia stopped exporting the material to Singapore (Channel News Asia, 2007).

Cost of agricultural land degradation (−)

The *Cost of agricultural land degradation* contributes only very marginally to the environmental items, because Singapore has very little agricultural production

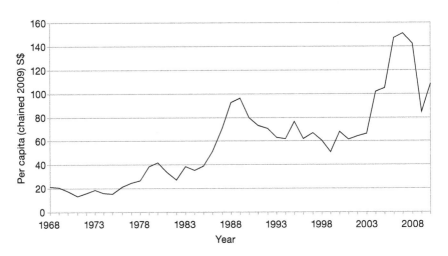

Figure 6.15 Cost of non-renewable resource depletion, Singapore

(Figure 6.16). Given the size of the country, Singapore has not been able to grow sufficient food to feed its population for a very long time. While virtually all the food consumed in Singapore is now imported, until the mid-1980s Singapore was able to produce a relatively large proportion of the meat, fish, eggs and vegetables consumed in the city-state. The most important farming activity (accounting for more than two thirds of the value of food produced) was rearing pigs and poultry. Domestic production met about 73 per cent of pork meat, 41 per cent of poultry meat, and 68 per cent of hen eggs consumed in Singapore (Khan, 1988: 179). In the late 1980s pig farming was phased out because of its polluting. By the late 1990s, the agricultural sector of Singapore was mainly engaged in the production of eggs, fish and vegetables for local consumption, as well as orchids and ornamental fish for export. The production of pork and chicken dropped considerably. In the second half of the 2000s about 300 pigs were slaughtered yearly, compared to some 870,000 in 1986 (Singapore Yearbook). At the same time, much agricultural land was set aside for housing and commercial buildings. In 1976, farmland accounted for 10,370 ha. By 1986 it had dropped to less than 4,000 ha. It is now approximately 1 per cent of the total landmass of Singapore, about 700 ha. The drop in production results in a drop of the *Cost of agricultural land degradation*, because this item is calculated based on the income from agricultural production (see Chapter 4). While in 1968 the *Cost of agricultural land degradation* contributed 1.2 per cent to the environmental items, by 2010 its contribution had dropped to 0.01 per cent.

Cost of fishery depletion (−)

The *Cost of fishery depletion* is calculated from the fish catch: the more fish is caught, the higher the cost. Figure 6.17 shows a considerable increase in the Cost

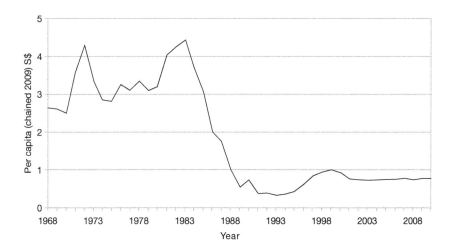

Figure 6.16 Cost of agricultural land degradation, Singapore

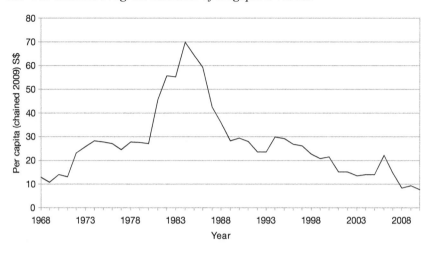

Figure 6.17 Cost of fishery depletion, Singapore

of fishery depletion, from S$13 per capita in 1968, to S$70 per capita in 1984, and subsequent drop due to the gradual end of fishing. From 25,041 tonnes caught in 1984, the catch dropped to 4,826 tonnes in 2010, representing a total decline of 81 per cent. The overwhelming majority of this fish comes from local aquaculture (AVA, 2014). Because of the way this item is calculated, a lower fish catch reduces the cost of fishery depletion, whatever the reason for the lower catch (past overfishing, pollution, alternative more attractive employment opportunities open to fishermen, or conservation efforts). Singaporeans are great consumers of fish, with fish consumption estimated at 100,000 tonnes per year, or 19 kg per person per year (AVA, 2014). However, most of the fish that is now consumed in Singapore originates from other countries, and the associated costs are not accounted for in Singapore's GPI.

Cost of air pollution (−)

In the early 1990s it was estimated that transportation was responsible for about 65 per cent of the air pollution in Singapore, power generation 25 per cent, and industry 10 per cent (Khan, 1994). To address road pollution (and road congestion) the government introduced a number of instruments, such as road pricing and a vehicle quota scheme (VQS) in the more congested areas, as well as legislative and fiscal measures to improve land use planning, the public transport system, road infrastructure, and traffic management (Chin, 1996; Koh, 2011).

These measures in the transportation sector were part of a broader set of policies to address air pollution and the inefficient use of energy. In April 2001, the government set up the National Energy Efficiency Committee (NEEC), initially to encourage businesses to use energy more efficiently. As such, the NEEC promoted the use of cleaner energy sources, such as natural gas and renewables (NEA, 2014), and positioned Singapore as a testing ground for pioneering energy technology. In 2006, following Singapore's ratification of the Kyoto Protocol, the NEEC was renamed National Climate Change Committee (NCCC). The NCCC's goal is that of addressing climate change by promoting greater energy efficiency, raising public awareness, and understanding the country's vulnerability to climate change (Hong, 2007).

These policies for a cleaner air were part of a broader approach to develop Singapore as a Green City. In 1992, Singapore introduced its first Green Plan (SGP), a master strategy to improve the environment with a vision to transform Singapore into a healthier and greener city by the year 2000. As the plan was implemented, government and non-government agencies monitored its progress and gathered public feedback. This led the way to The Singapore Green Plan 2012 issued in 2002 (MoE, 2002), a ten-year national blueprint to build a sustainable environment, which is updated every few years (Soh and Yuen, 2006). The 2006 update was dominated by the impact of particulates and climate change.

The *Cost of air pollution* (Figure 6.18) is relatively small in Singapore. Two spikes can be noticed, one in 1997 and the second in 2006. These anomalies were due to trans-boundary smoke hazes from forest fires in Indonesia, which consequently raised the respective level of PM_{10} particulates for both years (Koe *et al.*, 2001; Glover, 2006). The importance of the Indonesian haze in Singapore underlines the fact that tackling air pollution requires a collaborated effort of different countries, as neighbouring countries play a part in domestic conditions. The 2002 dated Singapore Green Plan 2012 highlights some of the international efforts made to combat air pollution (Chua, 2002). These include the signing of the ASEAN Agreement on Trans-boundary Haze Pollution in 2002, which sets out a legal framework for greater collaboration to prevent, monitor and mitigate haze pollution (Jerger, 2014).

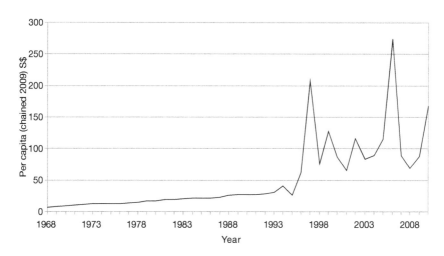

Figure 6.18 Cost of air pollution, Singapore

Cost of environmental degradation (−)

The *Cost of environmental degradation* is the second most important item (Figure 6.19). In Hong Kong, these costs are split up into categories of pollution (air, water, solid waste, and noise; see Chapter 5). The Singapore government does not split its expenditure into these categories, so this item cannot directly be compared to those of Hong Kong. While the various items are merged, we can gauge from The Singapore Green Plan 2012 (MoE, 2002) that the Singapore government emphasises solid waste management, ecology and biodiversity conservation, air pollution prevention, water pollution treatment, and the preservation of sources of fresh water.

The first focus area of the report is the ongoing task of handling solid waste in the confined spaces of Singapore. The daily output of solid waste in Singapore has increased considerably during the period under consideration, a reflection of Singapore's population growth, economic development, and structural transformations. From 1,260 tonnes of solid waste a day in 1970, Singapore produced 7,000 tonnes of solid waste a day in 2005. A growing amount of solid waste is recycled, from 40 per cent in 1998 to 56 per cent in 2008. However, this has not been able to keep pace with the increasing amount of waste generated (Zhang *et al.*, 2010). Most (91 per cent) of the waste collected is incinerated because of the limited land space available. The remaining 9 per cent, along with the ash generated from incineration, are disposed of at the Pulau Semakau landfill (NEA and MEWR, 2006).

The main island of Singapore once had two landfill sites, the Lim Chu Kang landfill in the north-western part of Singapore, which was filled in 1992, and the Lorong Halus landfill in the north-eastern part of Singapore, which was filled in 1999 (Bai and Sutanto, 2002). Presently, solid waste is disposed on Semakau

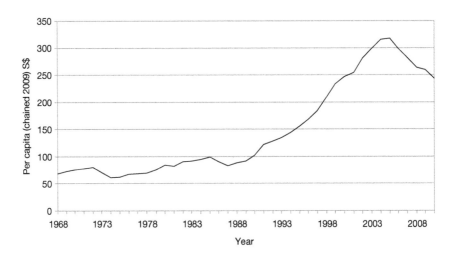

Figure 6.19 Cost of environmental degradation, Singapore

island, which has been transformed to meet Singapore's waste disposal needs, and is the country's only landfill site. The facility covers a total area of 350 ha and has a landfill capacity of 63 million m³. Semakau landfill is situated 25 km south of mainland Singapore (Khoo, 2009; Bai and Sutanto, 2002). Phase 1 is expected to be filled by 2019, Phase 2 by 2027 and Phase 3 by 2045 (Zhang *et al.*, 2010). In pursuit of a 'zero landfill' policy, the government hopes to extend the lifespan of the Landfill through the coordinated efforts of a national waste recycling programme, cooperation with the private sectors, such as the Waste Minimisation and Recycling Association of Singapore, tighter regulation and enforcement, and by means of tariffs and charges to reduce the amount of solid waste (Koh, 2011).

In terms of wastewater management, in 1977 Singapore conducted one of its most ambitious water management projects, which was critical in rejuvenating the city-state after independence, the ten-year water management project aimed at cleaning up the Singapore River and the Kallang Basin, which drains five other rivers (Chou, 1998). The process involved phasing out or relocating duck and pig farms within the catchment area, relocating boatyards from Kallang Basin, re-siting all 5,000 street hawkers to newly-built food centres built by the Housing Development Board, Urban Redevelopment Authority, and the Ministry of the Environment, and re-establishing biodiversity (Chou, 1998). Current methods of treating sewage include the Deep Tunnel Sewerage System, which includes an 80 km underground tunnel to transfer used water to one of six water-reclamation plants for treatment before discharging the water into the sea (Koh, 2011). Besides management projects, the government has also passed legislation to enforce and protect its water sources. For example, licenses are required for pig farmers to regulate effluent entering watercourses (Lim *et al.*, 2011; Koh, 2011).

Interestingly, the *Cost of environmental degradation* has been dropping from 2006 (Figure 6.19), which may reflect both an increasing quality of life necessitating fewer investments (perhaps as a result of better regulation or good results from previous investments), a decreasing interest by the government to alleviate environmental degradation resulting in decreasing expenditures, or an increasing population which reduces per capita costs.

Cost of climate change (−)

The largest environmental cost by far is that of climate change (Figure 6.20). Calculations are based on the abundance and global warming potential of each of the main greenhouse gasses: carbon dioxide (CO_2), methane (CH_4), nitrous dioxide (NO_2), hydrofluorocarbons (HFCs), perfluorocarbons (PFCs), and sulfur hexfluoride (SF_6). The per capita *Cost of climate change* grew rapidly, from $83 in 1968 to over $7,000 in the 2000s. This rapid increase is partly due to the fact that the cost is accumulated to emphasize that natural capital services cannot be substituted by man-made capital. We must point out that data for *Cost of climate change* were only available for the years 1990, 1995, 2000 and 2005, so the data used were extrapolated from these years. Per capita greenhouse

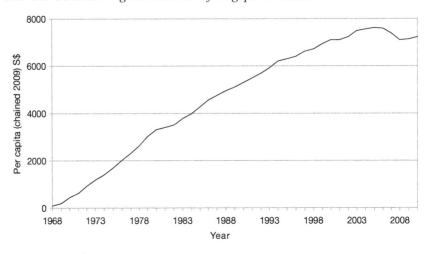

Figure 6.20 Cost of climate change, Singapore

gas emissions declined from 2006 to 2008 due to population growth. Total amount of emissions continued to rise.

Singapore is a non-Annex I country in the Kyoto Protocol, and therefore is not obligated to meet the emission reduction targets imposed on Annex I countries, even though its development status, in terms of level of wealth and per capita greenhouse gas emission levels, were in line with those of developed countries, even in the early 1990s when the Kyoto Protocol was negotiated (Hamilton-Hart, 2006). As with many non-Annex I countries that joined the Protocol, this has meant that Singapore has been able to maintain a business-as-usual mentality, continuing to depend on fossil fuels without pressure to meet emission targets. For its part, the government has introduced an efficiency-labelling scheme that targets certain household appliances in an attempt to reduce energy usage (ibid.). Despite this effort, the use of economic tools and emission caps would probably be more effective to influence manufacturers and consumers to reduce carbon emissions.

Cost of lost wetlands (−)

Singapore's rapid rate of economic development, accompanied by its small size, has resulted in the reclamation of land for new buildings. *Cost of lost wetlands* reflects the environmental costs related to the reduction of wetlands, and serves as a reminder of the importance of preserving the remaining mangroves (Figure 6.21). The remaining mangrove forests are located along the western and northern coasts of the main island and on several offshore islands. Most mangrove forests are essentially medium to small remnant patches (of up to 50 ha), and isolated strips (of up to 10 ha) confined to tidal rivers, river mouths, and sheltered bays, except for the mangroves on Pulau Tekong and Pulau Ubin (Hilton and Manning,

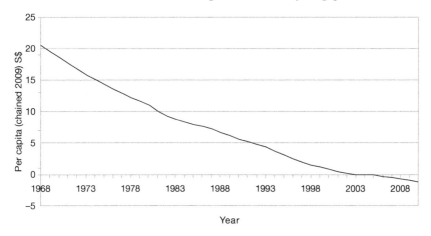

Figure 6.21 Cost of lost wetlands, Singapore

1995; Yang *et al.*, 2011). Yang *et al.* (2011) estimated that in 2010 mangrove forests covered 734.9 hectares, or about 1 per cent of Singapore's total area. By comparison, the estimated mangrove areas in Singapore were 6,400 hectares in 1953, 600 hectares in 1987 (Corlett, 1987), 483 ha in 1993 (Hilton and Manning, 1995), and 500 ha in 2005 (Basu, 2005). The increase recorded from 1993 to 2010 (reflected by a drop in costs in Figure 6.21) could be due to the use of different standards to estimate mangrove areas, the regeneration of disused prawn ponds, and continual reforestation and restoration efforts (Turner and Yong, 1999; Yang *et al.*, 2011).

One such effort to prevent further degradation of wetlands is the establishment of the Sungei Buloh Wetland Reserve. It was originally proposed to reclaim this area for high-tech agriculture in October 1987 (Francesch-Huidobro, 2008). However, counter-proposals from the Nature Society (Singapore), an environmental non-governmental organization, successfully averted the planned reclamation and let the area be designated a bird reserve (Francesch-Huidobro, 2008; Wee and Hale, 2008). In addition, in 2002 the wetland was eventually recognized as a site of international importance for migratory shorebirds (Francesch-Huidobro, 2008). This example helps highlight some of the efforts to conserve wetland areas in this small city-state with one of the world's highest population densities. The pressure is on, but Singapore's data indicates a welcoming trend that existing wetlands are being preserved and protected.

Environmental sub-index (−)

Figure 6.22 includes all the environmental items. All environmental items have a negative sign, as the environmental conditions have been deteriorating during the period under consideration. This is not surprising, since economic activities consist in the extraction and transformation of natural products, whose

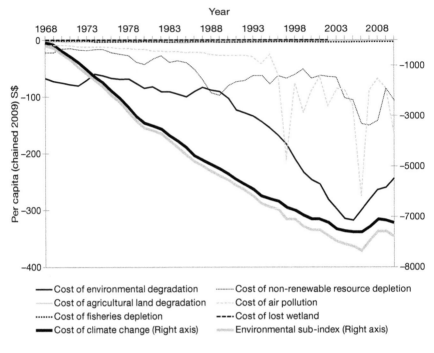

Figure 6.22 Environmental items of the Singapore GPI, Singapore

consumption, and eventual disposal also take a toll on the environment. Most costs in Figure 6.22 are rising with time, as the economic output increases, and the costs accumulate. The *Cost of climate change* is by far the largest, and is depicted on the right axis.

Despite rising costs, the city of Singapore has been promoting itself as the 'Garden City'. Tan *et al.* (2013) compared Singapore to other high-density cities, and found that the provision of urban green spaces in Singapore is not significantly differentiated from that of other high-density cities. However, Singapore has a high level of heterogeneity in the distribution of urban green spaces, and this may contribute to the impression that Singapore is greener than other cities in the region: 'the physical distribution of vegetation in the urban fabric is more important than the absolute quantum of vegetation to create a perception of pervasive greenery' (Tan *et al.*, 2013: 24). Urban planners have also sited cleaner and lighter industries closer to residential areas, while general and special industries (e.g. oil refineries, petrochemical complexes, pharmaceutical plants, and electronic industries) are located on industrial estates away from residential areas and water catchment areas, or on islands (NEA, 2009).

The GPI of Singapore: aggregate items

Figure 6.23 shows the aggregate data of the three different components of the GPI, together with the GPI and the GDP figures. As with Hong Kong, the GDP of Singapore increased considerably during the period under consideration, in an almost linear fashion, except for a few years of contraction. It can be said that Singapore has come a long way to reach the level of development and prosperity it currently enjoys, as the city-state has struggled with the transition from being a port under British sovereignty to being a host of labour-intensive manufacturing industries and finally to its current position as a centre of knowledge- and innovation-based activities. The GDP of Singapore increased even more than that of Hong Kong, going from S$7,030 in 1968 to S$62,191 in 2010, an average increase of 5.33 per cent a year for 43 years.[2] This growth has been the result of price stability, a high level of domestic savings, high net inflow of foreign direct investment, nearly full employment for most years, a strong positive balance of the current account, large foreign exchange reserves, a stable exchange rate system, and little external debt (Lim, 2010).

On the other hand, we can see that the GPI has not increased in the same fashion. During the 43 years under consideration, the GPI increased from S$4,695 to S$21,386, a yearly growth rate of 3.68 per cent, well below the growth rate of the GDP. During this period, the relatively high rate of growth of the economic subindex (5.17 per cent a year) was weighed down by the growing environmental and social costs, which increased by 8.91 per cent and 5.17 per cent (the same growth rate as the economic sub-index) a year, respectively. Clearly, social and

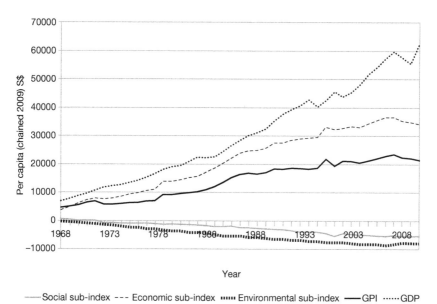

Figure 6.23 Comparison of the different sub-indices, the GPI and the GDP, Singapore

environmental costs weighed down the benefits from economic growth. However, there is a clear difference between the period from 1968 to 1993, 1993 to 1999, and 1999 to 2010 (Figure 6.23). For the 26 years from 1968 to 1993, the GPI grew by 5.61 per cent a year (similar to the growth rate of the GDP), to reach S$18,384 in 1993. This period saw a real improvement in the standard of living of the people, rather than simply an improvement in economic circumstances, at the expense of social and environmental conditions. However, by 2010 the GPI per capita had increased to only S$21,386, which means that the 18 years from 1993 to 2010 saw a growth rate of only 0.89 per cent a year. In 2010, the GPI was actually slightly lower than it was in 1999. From 1993 to 2010, while people's incomes and the amounts of market products they could purchase increased, this increase was largely outweighed by worsening social and environmental conditions. Ultimately, the quality of life, as measured by the GPI, hardly changed from 1993 to 2010. This finding is significant because it suggests that, although the economy (as measured by the GDP) increased considerably during this period, the social and environmental costs associated with economic growth almost completely wiped out the welfare benefits of that economic growth. Since 1993, the people would have been better off if the government had tried to improve their welfare, rather than increase economic output.

The lack of GPI growth since 1993 can also be traced to the fact that we estimate the GPI *per-capita*, and the population of Singapore has been increasing by 53 per cent from 1993 to 2010. In 1993, Singapore had a population of 3.31 million, and in 2010 it had a population of 5.08 million. This affects the per capita value of some items. Some may consider this approach to estimating welfare a weakness of the GPI. However, one may also argue that there is indeed a drop in welfare if the same roads are used by 53 per cent more cars, the MRT (the local subway system) has 53 per cent more passengers, or parks have 53 per cent more visitors. Rapid increase in the population may indeed worsen the standard of living, or welfare, of the population, and the drop of the GPI reflects this. At the same time, some per capita costs (e.g. *Cost of security and external relations*) drop when total expenditure remains unchanged while a country's population increases. So an increasing population may also contribute to higher per capita GPI figures.

These GDP values demonstrate Singapore's notorious economic growth, and reinforce its position as an Asian economic tiger. However, it is obvious that from the early 1990s, the gap between the performance of the GDP and that of the GPI has been growing, to the point where we can say that there is no relationship between the two. This increasing gap calls into question the use of GDP as a measure useful to policy-makers. Figure 6.24 represents the relationship between per capita GDP and per capita GPI, showing that initially both grew substantially, but that the positive relationship gradually decreased, and virtually disappeared as the per capita GDP continued increasing. This trend will be discussed further in the next chapter, where we discuss the threshold hypothesis.

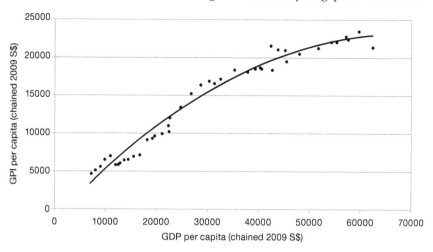

Figure 6.24 Per capita GPI versus per capita GDP: Singapore, 1968–2010

Conclusions

In this chapter we have seen that the GPI of Singapore has not increased throughout the study period. This is unlike the situation in Hong Kong, where the GPI, although slowing recently, and showing considerable irregularities, has continued growing during the period under consideration. The 1997 Asian Financial Crisis harmed both countries. However, Hong Kong seems to have better recovered. It also seems that Hong Kong has been able to benefit much more from its closeness to China, than Singapore has been able to benefit from its closeness to Malaysia and Indonesia.

In the case of Singapore, we have seen that the GDP was a fairly useful indicator of welfare until 1993, but since 1993 the relationship between GDP and GPI has virtually disappeared. The environmental and social costs have been rising more rapidly than the economic benefits, which has resulted in a rather 'flat' GPI since 1993, and in particular since 1999. This draws attention to the fact that while per capita GDP has roughly doubled from 1993 to 2010, the benefits in terms of welfare, as measured by the GPI, have been almost inexistent. This means that most of the efforts expended to grow the economy since 1993 have been wasted, or counter-productive. It would have been better to focus on addressing the environmental, social, and economic problems (e.g. inequality), rather than to attempt to further economic growth. This trend of a flat, or falling, GPI is common to many other countries whose economy has developed beyond a certain level. In the following chapter we discuss why this is so, and why it was not observed in Hong Kong.

Notes

1 The 16 domains are: housing, friendships, marriage, standard of living, household income, health, education, job, neighbours, public safety, condition of the environment, social welfare system, democratic system, family life, leisure, and spiritual life (Kuan *et al.*, 2010).
2 All figures in this discussion are in chained 2009 S$.

References

AVA. (2014). Aquaculture. Agri-Food & Veterinary Authority of Singapore. Retrieved 10 November 2014 from www.ava.gov.sg

Bai, R. and Sutanto, M. (2002). The practice and challenges of solid waste management in Singapore. *Waste Management*, 22, 557–567.

Basu, R. (2005). Mangrove appeal. The Straits Times, 1 March 2005.

Bernstein, S., Lerner, J. and Schoar, A. (2013). The investment strategies of sovereign wealth funds. *The Journal of Economic Perspectives*, 219–237.

Cahyadi, G., Kursten, B., Weiss, M. and Yang, G. (2004). Global Urban Development: Singapore Metropolitan Economic Strategy Report, Prague, Czech Republic: Global Urban Development.

Chan, K. B., Lai, G., Ko, Y. C. and Boey, K. W. (2000). Work stress among six professional groups: the Singapore experience. *Social Science & Medicine*, 50(10), 1415–1432.

Channel News Asia. (2007). Kekek Quarry reopening: Minimal disruption. Singapore: Channel News Asia.

Chin, A. T. (1996). Containing air pollution and traffic congestion: transport policy and the environment in Singapore. *Atmospheric Environment*, 30(5), 787–801.

Chiu, S. W. K., Ho, K. C. and Lui, T. L. (1997). *City-states in the global economy: industrial restructuring in Hong Kong and Singapore* (p. 75). Boulder, CO: Westview Press.

Chou, L. M. (1998). The cleaning of Singapore River and the Kallang Basin: approaches, methods, investments and benefits. *Ocean and Coastal Management*, 38, 133–145.

Chua, L. H. (2002). *The Singapore green plan 2012: Beyond clean and green towards environmental sustainability*. Ministry of the Environment.

Das, S. B. (2010). *Road to Recovery: Singapore's Journey Through the Global Crisis* (Vol. 413). Institute of Southeast Asian Studies.

Donaldson, J. A., Loh, J., Mudaliar, S., Md Kadir, M., Wu, B. and Yeoh, L. K. (2013). Measuring Poverty in Singapore: Frameworks for Consideration. *Social Space*, 58–66.

DOS. (2012). 'Household Income Increased in 2011 for All Income Groups' Singapore Department of Statistics. Press release, 14 February 2012.

Fields, G. S. (1994). Changing labor market conditions and economic development in Hong Kong, the Republic of Korea, Singapore, and Taiwan, China. *The World Bank Economic Review*, 8(3), 395–414.

Francesch-Huidobro, M. (2008). Governance, Politics and the Environment. Institute of Southeast Asia studies, Singapore.

Fuyuno, I. and Cyranoski, D. (2006). City state hopes research cash will buy global status. *Nature*, 442 (7099), 118–119.

Ganesan, N. (2005). *Realism and Interdependence in Singapore's Foreign Policy*. London: Routledge.

Glover, D. (2006). *Indonesia's fires and haze: the cost of catastrophe*. IDRC.

Hamilton-Hart, N. (2006). Singapore's Climate Change Policy: The limits of learning. *Contemporary Southeast Asia, 28*, 363–384.

Hilton, M. J. and Manning, S. S. (1995). Conversion of coastal habitats in Singapore: Indications of unsustainable development. *Environmental Conservation*, 22: 307–322.

Hong, M. (2007). Energy Perspectives on Singapore and the Region. Singapore: ISEAS.

Jerger Jr, D. B. (2014). Indonesia's role in realizing the goals of ASEAN's agreement on transboundary haze pollution. *Sustainable Development Law & Policy, 14*(1), 7.

Kee, H. L. and Hoon, H. T. (2005). Trade, capital accumulation and structural unemployment: an empirical study of the Singapore economy. *Journal of Development Economics, 77*(1), 125–152.

Keng, K. A., Kuan, T. S., Jiuan, T. S. and Kwon, J. (2004). *Understanding Singaporeans: Values, Lifestyles, Aspirations and Consumption Behaviors.* Singapore: World Scientific.

Khan, H. (1988). Role of agriculture in a city-state economy: the case of Singapore. *ASEAN Economic Bulletin, 5*(2): 178–182.

Khan, H. (1994). The strategies for sustainable development in a city-state economy: the case of Singapore. XXII International Conference of Agricultural Economists, Harare, Zimbabwe, 22–29 August 1994.

Khoo, H. H. (2009). Life cycle impact assessment of various waste convention technologies. *Waste Management*, 29, 1892–1900.

Koe, L. C., Arellano Jr, A. F. and McGregor, J. L. (2001). Investigating the haze transport from 1997 biomass burning in Southeast Asia: its impact upon Singapore. *Atmospheric Environment, 35*(15), 2723–2734.

Koh, K. L. (2011). Urban and industrial environmental management: The Singapore model. *Environmental Policy and Law, 41*(2), 102–103.

Krause, L. B. (1987). *The Singapore Economy Reconsidered.* Singapore: Institute of Southeast Asian Studies.

Kuan, T. S., Jiuan, T. S. and Keng, K. A. (2010). The wellbeing of Singaporeans. World Scientific, Singapore.

Li, K. W. (2002). *Capitalist Development and Economism in East Asia: The rise of Hong Kong, Singapore, Taiwan, and South Korea.* London: Routledge.

Lim, C. Y. (2010). Singapore Growth Model: Its strengths and its weaknesses. In: Sng, H. Y. and Chia, W. M. (Eds), *Singapore and Asia: Impact of the Global Financial Tsunami and other economic issues* (pp. 123–132). Singapore: World Scientific.

Lim, L. and Pang, E. (1986). *Trade, employment and industrialization in Singapore.* Geneva: International Labour Office.

Lim, M. H., Leong, Y. H., Tiew, K. N. and Seah, H. (2011). Urban stormwater harvesting: a valuable water resource of Singapore. *Water Practice & Technology, 6*(4).

MoE. (2002). The Singapore Green Plan 2012. Ministry of the Environment. Retrieved 10 October 2014 from http://app.mewr.gov.sg/data/ImgCont/1342/sgp2012.pdf

MoM. (2014). Foreign Workforce Numbers. Singapore: Ministry of Manpower. Accessed 14 November 2014 at www.mom.gov.sg/statistics-publications/others/statistics/Pages/ForeignWorkforceNumbers.aspx

NEA. (2009). Code of Practice on Pollution Control. Singapore: National Environment Agency. Retrieved 10 October 2014 from app2.nea.gov.sg/data/cmsresource/20090312534898283541.pdf

NEA. (2014). Air Quality and Targets. Singapore: National Environment Agency. Retrieved 10 October 2014 from http://app2.nea.gov.sg/anti-pollution-radiation-protection/air-pollution-control/air-quality-and-targets

NEA and MEWR. (2006). Integrated solid waste management in Singapore. National Environment Agency and Ministry of Environment and Water Resource. In: Asia 3R Conference, October 30–1 November, Singapore (accessed 00.06.06).

NEEI (National Energy Efficiency Initiative). (2009). Smart Grid, Smart City. Department of Environment, Water, Heritage and the Arts: Australia.

Ngiam, K. J. (2000). Coping with the Asian Financial Crisis: The Singapore Experience. *From crisis to recovery: East Asia rising again.*

Ong, E. and Tim, M. H. (2014). Singapore's 2011 General Elections and Beyond: Beating the PAP at Its Own Game. *Asian Survey, 54*(4), 749–772.

Peebles, G. and Wilson, P. (1996). *The Singapore Economy.* UK: Edward Elgar.

Poulgrain, G. (1998). *The Genesis of Konfrontasi: Malaysia, Brunei, Indonesia, 1945–1965.* C. Hurst & Co. Publishers.

Rajan, R. S. (2009). *Singapore: Trade, Investment and Economic Performance.* World Scientific.

Rossi, V. and Walker, J. (2005). *Assessing the Economic Impact and Costs of Flu Pandemics Originating in Asia.* Oxford: Oxford Economic Forecasting Ltd.

Shatkin, G. (2014). Reinterpreting the Meaning of the 'Singapore Model': State Capitalism and Urban Planning. *International Journal of Urban and Regional Research, 38*(1), 116–137.

SIPRI. (2014). SIPRI Military Expenditure Database. Retrieved 10 November 2014 from www.sipri.org/research/armaments/milex/milex_database

Soh, E. Y. and Yuen, B. (2006). Government-aided participation in planning Singapore. *Cities, 23*(1), 30–43.

Tan, A. T. (1999). Singapore's defence: capabilities, trends, and implications. *Contemporary Southeast Asia*, 451–474.

Tan, P. Y., Wang, J. and Sia, A. (2013). Perspectives on five decades of the urban greening of Singapore. *Cities, 32*, 24–32.

Tan, T. (2013). 'Singapore's Maids: No Respite?' *The Diplomat.* Retrieved 14 November from http://thediplomat.com/2013/08/singapores-maids-no-respite/

Temasek. (2014). Global Exposure. Singapore: Temasek. Retrieved 13 November 2014 from www.temasek.com.sg/portfolio/portfolio_highlights/geography

Thite, M., Wilkinson, A. and Shah, D. (2012). Internationalization and HRM strategies across subsidiaries in multinational corporations from emerging economies – A conceptual framework. *Journal of World Business, 47*(2), 251–258.

Turnbull, C. M. (2009). *A History of Modern Singapore, 1819–2005* (Revised Edition). Singapore: NUS Press.

Turner, I. M. and Yong, J. W. H. (1999). The coastal vegetation of Singapore. In: Briffett C. and Ho H. C. (Eds) *State of the Natural Environment in Singapore* (pp. 5–23). Singapore: Nature Society (Singapore).

Van Dyne, L. and Ang, S. (1998). Organizational citizenship behavior of contingent workers in Singapore. *Academy of Management Journal, 41*(6), 692–703.

Wai-lam, C. (2000). Unemployment-related benefits systems in Singapore. Singapore: Research and Library Services Division Legislative Council Secretariat.

Watkins, K. (2007). Human Development Report 2007/2008: fighting climate change.

Wee, Y. C. and Hale, R. (2008). The Nature Society (Singapore) and the struggle to conserve Singapore's nature areas. *Nature in Singapore, 1*, 41–49.

Wong, C. M. and Foo, K. H. (2011). Motivational functions, gender, age and religiosity influences on volunteerism: a Singapore volunteer organisation perspective. *Journal of Tropical Psychology*, 1(01), 31–44.

Wong, P. K. (2002). From using to creating technology: the evolution of Singapore's national innovation system and the changing role of public policy, in Lall, S., and Urata, S. (Eds). Foreign Direct Investment, Technology Development and Competitiveness in East Asia. Singapore: Edward Elgar.

Yang, S., Li, R. L. F., Sheue, C.-R., Yong, J. W. H. (2011). The Current Status of Mangrove Forests in Singapore. Proceedings of Nature Society, Singapore's Conference on 'Nature Conservation for a Sustainable Singapore' – 16th October 2011, pp. 99–120.

Yeoh, F. (2006). National research foundation of Singapore. Paper presented at the ICAAS-UIAAS Innovation Symposium, Singapore, 26 August 2006.

Yeung, H. W. C., Poon, J. and Perry, M. (2001). Towards a regional strategy: The role of regional headquarters of foreign firms in Singapore. *Urban Studies*, *38*(1), 157–183.

Zhang, D., Keat, T. S. and Gersberg, R. M. (2010). A comparison of municipal solid waste management in Berlin and Singapore. *Waste management*, *30*(5), 921–933.

7 The 'threshold hypothesis' and the two city-states

Introduction

The analysis of Chapters 5 and 6 has shown that in Hong Kong there has been a relatively close relationship between the GDP and the GPI during much of the period under consideration, while in Singapore that close relationship lasted only from 1968 to the late 1980s. In this chapter we discuss the reasons for such differences between Hong Kong and Singapore, and introduce the concept of 'threshold', the point at which further economic growth results in a gradual drop of the GPI, because the social and environmental costs outweigh the economic benefits. We start with an introduction of the threshold hypothesis and look at the reasons and consequences of such threshold in other countries. We then look at the situation of Hong Kong and Singapore and compare the two city-states.[1] In particular, we discuss why Hong Kong's GPI has not experienced such threshold.

The threshold hypothesis: other countries' perspectives

Before we compare Hong Kong to Singapore, it is useful to briefly review the GPI studies of other countries. Kubiszewski *et al*. (2013) compared the GDP and GPI of 17 different countries in Asia (China, India, Japan, Thailand, Vietnam), Europe (Austria, Belgium, Germany, Italy, Netherlands, Poland, Sweden, United Kingdom), Oceania (Australia and New Zealand), North America (United States), and South America (Chile). Together these 17 countries represent 53 per cent of the world population, and in 2008 about 60 per cent of world GDP. It seems fair to say that these data allow us to make some generalizations.

Figure 7.1 shows the GDP per capita of these countries from 1950 to 2008. As can be seen, the GDP per capita of most countries has steadily increased over the six decades under consideration, though in some cases with a small drop in the early 1990s. During these six decades, many countries' GDP has at least tripled. Looking only at economic output, as measured from the GDP, we can conclude that the post-WWII era has been very prosperous, as economic growth has continued throughout the period.

The GPI figures of these same 17 countries give a different picture (Figure 7.2).[2] Not all of these studies follow the same approach. Some use different items,

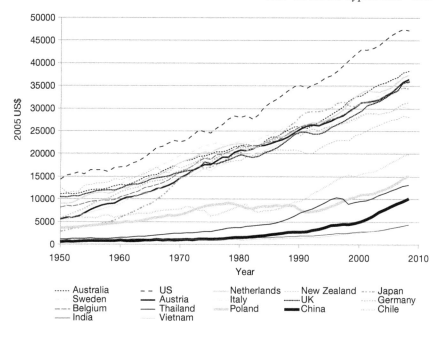

Figure 7.1 GDP per capita of selected countries (1950–2008)

Source: Adapted from Maddison (www.ggdc.net/MADDISON/oriindex.htm) in Kubiszewski *et al.* (2013)

and some estimate the GPI while others estimate the ISEW (which we introduced in Chapter 2). However, since the two indicators are similar and there are few methodological differences between the different studies, Kubiszewski *et al.* (2013) consider them sufficiently uniform for the present analysis. For most countries, the GPI has not been estimated for the whole period under consideration, but only for three to four decades, either from the 1950s to the 1990s, or from the 1960s to the 2000s. The data are converted to 2005 US$, by adjusting for currency and inflation, using Purchasing Power Parity.

It is immediately apparent that during these years the trend is not the same as that of the GDP. For most countries there is a general trend of a steady, consistent, increase in the GPI up to a point, after which countries' GPI either drop (in the case of the Netherlands for example), or remain flat (in the case of Austria for example). Only in a few cases does the GPI remain relatively flat (China, Vietnam), or grow (Japan, Thailand, Italy) from the beginning to the end of the study period. The phenomenon of a dropping GPI beyond a certain level of economic growth has already been identified in 1995 by Max-Neef. Max-Neef (1995) hypothesized that: 'for every society there seems to be a period in which economic growth (as conventionally measured) brings about an improvement in the quality of life, but only up to a point – the threshold point – beyond which, if there is more economic growth, quality of life may begin to deteriorate' (p. 117).

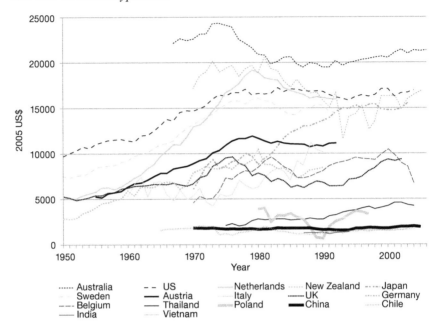

Figure 7.2 GPI per capita of selected countries (1950–2006)

Source: Adapted from Kubiszewski *et al.* (2013)

We can see that this threshold exists for the overwhelming majority of countries, but that the year in which the threshold was reached varies among countries. The first country to experience a drop in GPI was Australia, in 1974. The United Kingdom followed in 1976, the Netherlands and Austria in 1980, and most other countries in the 1980s. Some countries have since reversed the trend, and the UK for example has had a growing GPI from 1994 to 2002.

Kubiszewski *et al.* (2013) synthesized the GDP and GPI figures of the 17 countries to derive a global GPI (Figure 7.3). As can be seen from Figure 7.3, the GDP has been growing in an almost linear fashion from 1950 to 2003. By 2003 the GDP per capita was about three times larger than it was in 1950. On the other hand, the GPI started to drop in 1979, never to recover.

The decline in the GPI since 1979 means that our well-being, as measured by the GPI, has been slightly declining. Overall, the decline is not drastic, although for some countries (such as Australia and the Netherlands) it is quite considerable. However, the gap between per capita GDP and per capita GPI continued increasing, and by 2003 it was substantial. We can conclude that much of the efforts expended to grow the economy since 1978, not only has not made us better off, but it was even accompanied by a drop of welfare. It would have been better if government had attempted to increase welfare, by addressing such things as environmental degradation and inequality, rather than focusing on economic growth.

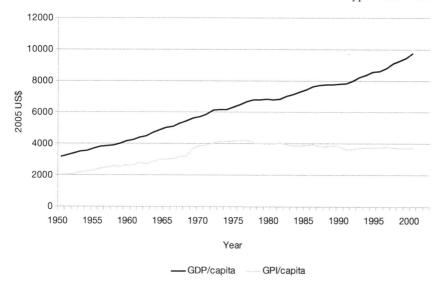

Figure 7.3 Adjusted global GPI per capita and GDP per capita[3]

Source: Adapted from Kubiszewski *et al.* (2013)

Why is there a threshold?

The reasons for a threshold may be multiple, depending on the country under consideration.

Growing social and environmental costs accompanying economic growth

Kubiszewski *et al.* (2013) use the same dataset of 17 countries' GPI and GDP figures to calculate the per capita GDP level at which per capita GPI starts to decline (Figure 7.4). Kubiszewski *et al.* (2013) estimate that until US$7,000 GDP per capita, the GPI per capita and the GDP per capita are highly correlated ($R2=0.98$). This is consistent with studies of subjective well-being, which level off after around US$7,000 per capita (Deaton, 2008; Inglehart, 1997). When the GDP grows beyond about US$7,000 per capita, there is a negative correlation ($R=0.61$) between GDP growth and GPI growth, with further economic growth lowering GPI as the social and environmental costs of economic growth outweigh the economic benefits.

These conclusions are made from averaging data from the 17 countries under consideration, and masking the great differences that exist among countries. The data also mask the fact that poorer countries have great difficulties to increase their GPI, which starts dropping much sooner as GDP grows (see below). So, a US$7,000 threshold is actually a conservative one. Poor countries' per capita GPI starts declining when GDP figures are much lower.

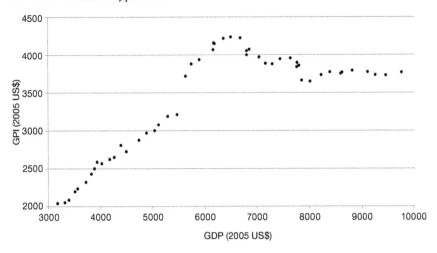

Figure 7.4 Global GDP/capita versus estimated global GPI/capita (all data in 2005 US$)

Source: Adapted from Kubiszewski *et al.* (2013)

These findings strengthen the argument that the GDP is actually a very poor indicator of welfare once GDP per capita reaches approximately US$7,000 (in 2005 PPP US$). This suggests that the marginal benefits yielded from rapid economic growth are greatly outpaced by increasing marginal costs in a finite world, as indicated by the law of diminishing marginal benefits and the first and second laws of thermodynamics (Lawn and Clarke, 2010). In other words, as the economy grows beyond a particular point, it is likely that the social and environmental costs associated with further economic growth grow faster than the benefits gained from additional economic output.

Deregulation

In the 1980s there was a push towards deregulation, privatization and weakening social welfare, first in the US and UK, and then in much of the world. It is likely that this contributed to an increase in the social and environmental costs, as well as inequality, although further research is needed to confirm the exact impact of that deregulation.

Free trade in a closed world

For all the countries that have displayed a threshold (these exclude Japan, Italy, Thailand, China and Vietnam, whose GPI does not drop), that threshold was reached within 12 years (between 1971 and 1982), and for eight countries in three continents within five years (between 1978 and 1982) (Table 7.1). It is interesting that so many countries have reached a threshold within such a relatively short period of time. In our opinion, the reason is related to the structure of the world

Table 7.1 Year in which the threshold was reached.

1971	Chile	
1972		
1973		
1974	Australia	
1975		
1976	United Kingdom	
1977		
1978	USA	
1979	Netherlands	Austria
1980	Sweden	Germany
1981	New Zealand	Poland
1982	Belgium*	

Notes: * Belgium's GPI grew again from 1987, and by 1999 it returned to the level it was in 1982

economy. Since World War II, trade barriers have weakened, and the world economy has integrated both globally (for example through such agreements as the GATT trade rounds, which culminated in the WTO), and regionally (for example through the actions of political and economic organization such as the Association of Southeast Asian Nations (ASEAN) and the Asia-Pacific Economic Cooperation (APEC)). This has resulted in a world that is much more closely integrated, with global booms and busts, and global prices for natural resources and manufactured products. The close integration of the world economy facilitates the export of labour-intensive manufacturing products and raw materials from poor to rich countries, and indeed, these are the sectors poor countries have specialized in. This specialization is usually argued to be to the benefit of poor countries. However, the same integration and specialization also have negative consequences for the development of poor countries. This is for two reasons.

Model of development pursued by poor countries

First, it is due to the model of development that poor countries are forced, or chose, to pursue. Many poor countries focus on a combination of: 1) extracting natural resources; 2) producing labour-intensive manufactured products; and 3) lowering environmental standards to attract polluting manufacturing plants. The extraction of natural resources, whether timber or mineral resources, is an easy way to raise GDP, and the profits from their sale can be used to develop the country. However, the exploitation of natural resources in many cases results in lower welfare, because of the environmental costs associated with the activity. For example, mining often results in air and water pollution, while logging results in soil erosion and is accompanied by social costs to the communities that rely on the forests for livelihood. Because of the high externalities, the economic benefits may as well be limited. Similarly, the development of labour-intensive manufacturing industries may make a contribution to the economy, but poor countries compete to attract labour-intensive industries by lowering labour standards and salaries, so the contribution is limited. Finally, raising environmental standards in rich countries has prompted companies to export their more polluting industries. Again, poor countries compete for foreign investment by lowering their own environmental standards, which has contributed to the low environmental standards that can now be found in some poor countries (Pao and Tsai, 2011; Knox *et al.*, 2014).

Overall, the benefits of that pattern of development are low. While the GDP increases, the GPI captures the environmental problems and limited social benefits that result from such pattern of development, and may as well report a drop in welfare. Furthermore, the profits of investors are often either exported to rich countries, or spent on luxury goods imported from rich countries, so even the economic benefits are limited for the overwhelming majority of the population. The result is that exporting natural resources, or producing labour-intensive products or products that generate much pollution, is not necessarily the best way to develop a country.

For rich countries, we have the opposite scenario: while the more environmentally degrading and labour-intensive industries are exported, the better paid, less polluting capital intensive manufacturing jobs are retained in rich countries. This helps them achieve a higher GPI than would otherwise be the case. This will be further discussed below, as this argument can explain the success of Hong Kong: the GPI of Hong Kong has kept growing through the years, because of the export of the more polluting, labour-intensive industries to China.

This pattern can be extrapolated from Figure 7.1. As can be seen in Figure 7.1, the economy of the six poorest countries has been growing steadily and by the early 2000s they had an average GDP of US$12,000 per capita (unweighted average), about 35 per cent as much as that of the 11 richest countries. On the other hand, Figure 7.2 shows that in the early 2000s the same six countries had a GPI of US$2,600 per capita (unweighted average), which is less than 20 per cent of the GPI of the richest 11 countries. The difference between poor countries' and rich countries' per capita GPI keeps increasing, even when the differences between poor countries' and rich countries' per capita GDP keeps decreasing. It seems that poor countries find it difficult to raise their citizens' welfare even as their economy grows, and that this is because free trade and the integration of the world economy gives them limited options of development. The world economy is organized in such a way that poor countries are unable to increase their citizens' welfare.

Competition for natural resources

Second, economic development and the integration of the national economies has contributed to a worldwide competition for natural resources, which prevents poor countries using them for their own development. The high demand in rich countries, and high prices that rich countries are able to pay for the raw materials, prevents poor countries from going through the same phases of development the rich countries did during the nineteenth and early twentieth centuries. That development entailed exploiting natural resources from the rest of the world. During that period, natural resources were more plentiful and cheaper, and there could be more waste. Environmental degradation did not have as many negative repercussions as it does now, partly because there were fewer people (in 1950 the world population was 2.5 billion). Lawn and Clarke (2008) have described this as rich countries having a 'headstart' in economic progress, when the cost associated with this development was offset to some extent by operating in an 'emptier' world, where there was less regulation, less competition and more scope for enterprise. In other words, the marginal costs did not meet or overtake the marginal benefits when the world had more 'quotas' – more opportunities and more leeway to absorb the exploitation and depletion of natural capital.

When poor countries started to develop their own economies in earnest, often after WWII, they increasingly have had to compete for the world's natural resources with rich countries. Rich countries can afford to pay more for natural resources because their citizens have higher incomes, and their industries are able to use these resources more efficiently (Lawn and Clarke, 2008). While some poor countries are able to export these raw materials and invest the proceeds, those poor countries that do not have a sufficiently large number of raw materials are virtually priced out of development.

Furthermore, from an environmental perspective there is the problem that pollution and the exploitation of natural resources by rich countries is already at levels that are detrimental to poor countries' sustainable welfare, for example because of global warming. All these things lower the threshold and 'stunt' poor countries' GPI growth. Overall, it has led to a contraction in the global threshold beyond which further growth is very difficult. It means that the 'latecomers' would reach the threshold point at a much lower level of sustainable welfare (i.e. lower per capita GPI) than developed countries did (Lawn and Clarke, 2010).

This can explain why countries have experienced a threshold at a similar time, regardless of their level of development, their GDP, or their GPI per capita: the world economy has integrated for all countries at approximately the same time, and all countries are now faced with the same constraints. For example, as goods are gradually becoming scarcer after decades of growth, their relative scarcity has led to a worldwide price hike that affects all countries simultaneously.[4] This implies that pushing for increasing economic growth in rich countries, not only does not lead to a growing welfare in these countries, so is not useful, but it also prevents poorer countries from improving the welfare of their citizens. This is contrary to mainstream economists' view that economic growth of rich countries' benefits (trickles down to) poor countries.

The GDP has been able to continue growing because more sophisticated technology allows people to extract increasingly scarce raw materials. Because of improved technology the prices of those raw materials do not necessarily increase when resources become scarcer (even though economic theory says they should). However, the extraction of scarcer resources is often accompanied with additional environmental degradation, for example because more energy is used to extract, more soil is removed, or the materials are transported over large distances. Thus, there might be an increase in the GDP, which either ignores environmental degradation or considers it a benefit to the economy, while the same extraction of scarcer resources results in a drop in the GPI, which accounts for these things negatively.

We can draw the following conclusion: the reason poorer countries cannot develop is the close integration of the world economy, and the high rates of consumption of rich countries. This has serious implications for the domestic and international policy outlook of developed countries. These countries should be considering whether to moderate their current level of economic growth (which does not even increase welfare), or else risk holding back developing countries from ever reaching the level of welfare currently enjoyed in developed countries (Lawn and Clarke, 2010). 'The ability of poor nations to increase their economic welfare may now be dependent upon rich countries abandoning their sole policy focus on GDP growth. This would provide the "ecological space" for poor nations to experience a phase of welfare-increasing growth' (Kubiszewski *et al.*, 2013: 66). If the GDP of rich countries continues to rise at the expense of poorer countries, then this could potentially affect political, environmental, social and economic stability worldwide, and for example lead to large-scale migration.

The case of Singapore and Hong Kong

In the following pages we compare the GPI of Hong Kong and Singapore, and discuss why Singapore seems to have reached a threshold, while Hong Kong not. We also discuss the 'Pollution Haven Hypothesis', to make further sense of the situation of the two city-states.

Hong Kong and Singapore are two city-states that have experienced similar circumstances and followed a similar development post-World War II. The post-WWII period ushered in an era that saw the two city-states develop to become two of the best performing economies in East/Southeast Asia. In spite of their historical similarities, the results of the GPI analysis shows some important differences. Figure 7.5 compares the GDP and GPI of Hong Kong and Singapore from 1968 to 2010. To facilitate comparison we use index values, whereby 1968 is given a value of 100. Between 1968 and 2010 Hong Kong's GDP per capita grew by an average rate of 4.54 per cent per annum. On the other hand, in the same period, GPI per capita grew by an average rate of 3.78 per cent per annum. Even though the GPI has stalled a few times (from 2000 to 2005, and again from 2007 to 2010), overall it has continued to grow throughout the 43 years under consideration. The same can't be said about Singapore. The GDP of Singapore increased even more than that of Hong Kong, at an average rate of 5.33 per cent a year for 43 years. Initially, the GPI of Singapore also increased very rapidly. For the 26 years between 1968 and 1993, the GPI grew at a rate of 5.61 per cent a year (very similar to the growth rate of the GDP). However, for the 18 years from 1993 to 2010 the GPI grew by an annual rate of only 0.89 per cent. Overall, over the 43 years under consideration the GPI of the two city-states has risen by a similar percentage, but in

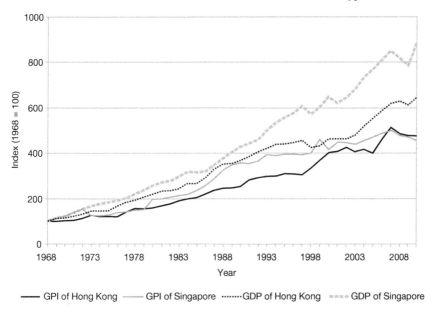

Figure 7.5 Comparison between the index values of Hong Kong's and Singapore's GDP and GPI

Hong Kong this rise has been more constant. In Singapore most of the growth occurred during the first 25 years. Time will tell whether 2007 was the threshold for both city-states, or simply a temporary drop in the GPI due to the financial crisis of 2007–2009 and the subsequent Great Recession. However, we can conclude that even though neither Hong Kong nor Singapore have experienced the threshold of most other countries (Figure 7.2 and Figure 7.4), the growth rate of Singapore's GPI has slowed considerable since 1993.

The GDP figures are unlike the GPI figures. The GDP has grown considerably more in Singapore than in Hong Kong. It seems that Hong Kong has been more successful in increasing the welfare of its population, while Singapore has been more successful in increasing the economic output, as measured by the GDP, but less so in increasing welfare, as measured by the GPI.[5] In the following pages we try to explain the differences between Singapore, Hong Kong, and other countries, in particular in terms of a lack of clear threshold.

Economic restructuring and deindustrialization

First and foremost, both Hong Kong and Singapore have experienced considerable economic restructuring and deindustrialization over the last 20 years. Their manufacturing sector has moved to other countries, while they have retained the higher paying (and cleaner) office jobs. Hong Kong has been able to do so because of its proximity and favoured relationship with China. Singapore has been able to do so because it is politically stable, has invested considerable sums in educating its workforce, and is located in the middle of a region that has striven to attract labour-intensive manufacturing industries, but did not have the educated workforce and political stability necessary to attract the regional headquarters of the multinational corporations that produce or market these goods.

Three events in particular helped Hong Kong businesses to expand in neighbouring China, or otherwise benefit from the political and economic transformations in the neighbouring Chinese provinces. First, in 1978, Deng Xiaoping visited Guangdong, the Chinese province that borders Hong Kong, and promoted the 'open door policy', whereby foreign investment was encouraged in China. Much of that investment initially came from Hong Kong. Second, in 1997 Hong Kong returned to China under the 'one country, two systems' model, which allowed Hong Kong to retain its judicial system (modelled on the British) and its ease to do business,[6] while it was given additional benefits by China, as a Chinese Special Administrative Region (SAR). Third, in 2001, China joined the WTO, which led to its rapid economic growth.

All these events encouraged and gradually facilitated Hong Kong businesses to relocate their manufacturing plants from Hong Kong to China, given the latter's lower wages and less stringent workplace laws and environmental regulations. While the low-wage jobs in the manufacturing sector moved to China, Hong Kong retained the better paid administrative jobs, resulting in higher wages and personal consumption expenditures. Even when investment in Guangdong (now called the 'manufacturing capital of the world') does not come from Hong Kong, Hong Kong inevitably benefits, as many of the goods produced transit through Hong Kong ports, and some of the newly rich people in Guangdong province shop in Hong Kong.

The relocation of industries to neighbouring China reduced the air and water pollution in Hong Kong, reducing the costs in the Environmental sub-index. China has become the pollution haven for Hong Kong manufacturing industries, while Hong Kong continues to develop the tertiary sector to service these industries. Some of these costs return to Hong Kong in the form of air pollution for example – but only some. The majority of the environmental costs of manufacturing (such as the pollution they generate) are accounted for in China, rather than Hong Kong. This contributes to the fact that China's GPI has not been able to improve much.

These two factors – the deindustrialization and retention of the better jobs in the service sector, and the fact that pollution is now generated in China – should go a long way towards explaining Hong Kong's continued growth in both per capita GPI and per capita GDP. Through its very rapid deindustrialization, it is quite possible that Hong Kong has been able to delay the threshold for longer than most countries are able to. On the other hand, while additional investment of Hong Kong businessmen in China may continue, most manufacturing plants have already relocated, and it is not obviously clear how Hong Kong may benefit from further economic and political reforms in China. This may also go some way towards explaining the drop in the GPI during the last years.

Similar arguments can of course also be made for Singapore. Singapore is famous for hosting the headquarter of multinational corporations with manufacturing plants in neighbouring countries, thus maintaining the high-paid administrative jobs, but not the low-paid jobs in the plants. As such, it is somehow similar to Hong Kong. However, in addition the Singapore government also tried to retain the city-state manufacturing sector, partly through foreign direct investment. In 2013, the industrial sector still made up 29.4 per cent of GDP, while in Hong Kong it made up only 6.9 per cent of GDP. Although much of Singapore's industrial sector consists in higher value-added electronics, biotechnology and pharmaceuticals, Singapore has not been able – or did not want – to completely shake off the more labour-intensive industries that carry less value added (WSJ, 2014). Of course having a strong industrial base is an asset in many regards, including economic resilience, but when such industry has a relatively large focus on labour-intensive manufacturing, there are fewer benefits, as measured by the GPI. Neumayer (2001) and Dong *et al.* (2012) have argued that in an effort to attract foreign investment, most countries accept environmental standards that are below their efficiency levels, or even fail to enforce such standards. While Singapore usually markets itself as a 'green city', it is likely that there is an extra burden carried in the form of environmental costs. Indeed, Singapore's GPI results show that environmental costs have grown faster than those of Hong Kong's GPI.

Expansion of neighbouring economies

One difference between Singapore and Hong Kong is that Singapore competes with its neighbours for foreign investment, while China has been happy to facilitate the growth of Hong Kong thorough favourable treatments, in particular in the financial sector. Also, Hong Kong is able to benefit from the expanding Chinese economy by attracting high-spending mainland tourists and investors (for example in the housing market), something that Singapore is less able to do. The construction of a casino on the island of Sentosa in Singapore (opened in 2010) may go some way towards trying to benefit from the economic expansion that took place in the neighbouring countries and China.

Lack of subsidies for particular sectors

Most large countries have particular sectors that they subsidize. In particular, the agricultural sector is often subsidized, either for political reasons (the workers are particularly well organized, or the agricultural sector may be particularly important in some provinces within federal governments, so they are able to have disproportionate power), for cultural reasons (the sector may be deemed important and contribute to national identity), or for security reasons (it may be deemed important to produce a minimal amount of food in the country). In 2009, the European Union paid US$120.8bn in farm subsidies, Japan US$46.5bn, the US US$30.6bn, Turkey US$22.6bn, and South Korea US$17.5bn (Agrimoney, 2010). Another sector that may be subsidized with little benefit to welfare is the defence. On the other hand, some countries subsidize the consumption of particular products, such as fossil fuels, which may add to welfare but is a drain on the economy. In other cases, governments introduce trade barriers to protect national industries that cannot compete with foreign competitors. All these 'inefficient' (though entirely justifiable) uses of capital may go some way to explain the drop in GPI of most countries in Figure 7.2. Subsidies are not as common in Hong Kong and Singapore, where sectors or industries that cannot survive without subsidies or trade barriers are left to decline, leading to a more efficient allocation of resources, lower taxation (which results in higher personal consumption expenditure) and more funds available to the government for investment, all of which increase the GPI.

Efficient use of public investment

Hong Kong and Singapore are relatively small city-states. In 2010, Hong Kong had 7 million people on 1,104 km², and Singapore had 5 million people on 718 km². Both countries have a very high population density, at 6,340 and 6,964 people per km², respectively. This can be compared to Taiwan, which has a population of 23 million people on 36,000 km², or a population density of 639 people per km², approximately ten times less. With such a large population density, government investment in infrastructure is more efficient, which results in higher *Services yielded from fixed capital*, and again allows for lower taxation, which leads to higher *Personal consumption expenditure*, among others. Also, the smaller size leads to lower *Non-renewable resource depletion* because transportation takes place over a shorter distance and public transportation can be more cheaply and efficiently organized, leaving again more funds for *Personal consumption expenditure*.

The GPI (of city-states) ignores imported goods

Hong Kong and Singapore are city-states whose economy focuses on the service sectors (in 2013 the service sector accounted for 70.6 per cent of the GDP of Singapore, and 93 per cent of the GDP of Hong Kong). The overwhelming majority of the food, manufactured products, energy, and raw materials they consume are imported. The agricultural and manufacturing sectors in particular generate considerable environmental degradation, increasing the environmental costs of the GPI. However, these costs are not accounted for in the cities' GPI. In addition, in many cases the agricultural and labour-intensive manufacturing sectors do not produce much value added, so do not generate a large *Personal and public consumption expenditure*. For this reason, it may be argued that the GPI of city-states whose economy is predominantly made up of the service sector may not be directly comparable to that of large countries with important agricultural and industrial sectors.

The same weaknesses arise when estimating the GPI of countries that are heavy importers of raw materials. This is the case of Japan for example. Japan imports most of its raw materials from other countries due to its relative lack of natural resources, and in turn exports large quantities of high-end manufactured products and services. The environmental costs associated with imports are accounted for in the countries that export these primary products, as is evident in the drastic decline in its environmental costs from 1980 to 2003 (Makino, 2008; Lawn and Clarke, 2008). This may go some way towards explaining why Japan has not experienced a 'threshold' point, in spite of its high per capita GDP.

One could argue that the environmental degradation caused by the manufacture of goods to be exported should be accounted for in the countries that import the goods, rather than the country that manufactures them. If the environmental and social costs of the goods imported by Hong Kong and Singapore were included in the city-states' GPI, it is likely that they would have experienced a threshold several years ago. As Clarke and Lawn (2008: 579) wrote about the state of Victoria in Australia, Hong Kong's increasing GPI may not be due 'to it operating more sustainably, but because it can "off-load" or "free-ride" on the unsustainable activities of other regions'.

Conclusions

In this chapter we have discussed the threshold hypothesis, and why it does not apply to Hong Kong and Singapore. The threshold hypothesis is a very important concept, as it shows the difficulties that countries have when trying to elevate the level of welfare beyond a certain point, even though the economic output keeps growing. This calls into question the strategy of promoting endless economic growth at any (environmental and social) costs, for example by weakening environmental standards or reducing the minimum wage. An important function of the GPI is to identify the point beyond which additional economic growth is no longer beneficial to the people. Beyond the threshold, countries should not strive for further economic growth, but instead should try to address the problems that cause a drop in welfare. How to do this should be the focus of national debate, as different options may be suitable for different societies. We will discuss some options in Chapter 8, where we review the Steady State Economy.

In this chapter we have also discussed why Hong Kong and Singapore do not display that threshold. The argument may be summarized as: 1) Not yet: the two city-states have been able to benefit from their particular geographic location, and the development that has occurred in the neighbouring countries. This is particularly the case for Hong Kong. However, it is likely that most of the benefits have already been realized, and the city-states cannot look forward to many more additional benefits. 2) The environmental degradation that accompanies the production of resource consumed in the city-states is accounted for in other countries. 3) The small size of the city-states allows for a more efficient use of domestic resources. This calls into question whether the GPI is a suitable tool to measure the welfare of city-states. Indeed, Clarke and Lawn (2008), and Bagstad and Ceroni (2007) discuss the weaknesses of applying the GPI at the city level, and in particular the fact that it is difficult to obtain data to estimate the GPI. One of the advantages of doing a GPI analysis of Hong Kong and Singapore, is that while in many respects they are cities with close economic relations with their 'hinterland' (in the case of Hong Kong, the Chinese province of Guangdong where many Hong Kong-owned factories operate, in the case of Singapore, neighbouring Malaysia and to some extent Indonesia), unlike national cities they maintain a system of accounting that is similar to the United Nations' sponsored System of National Accounts. Thus, the GPI can be estimated with the level of accuracy of a nation.

Notes

1   Hong Kong is of course not a city-state but a Special Administrative Region (SAR) within the People's Republic of China (PRC). We consider it a city-state here because of the considerable independence it has.
2   The GPI figures were estimated by Wen *et al.* (2008) for China, Lawn (2008b) for India, Makino (2008) for Japan, Clarke and Shaw (2008) for Thailand, Hong *et al.* (2008) for Vietnam, Stockhammer *et al.* (1997) for Austria, Bleys (2008) for Belgium, Rosenberg et al. (1995) for Germany, Italy, and the Netherlands, Gil and Sleszynski (2003) for Poland, Stymne and Jackson (2000) for Sweden, Jackson et al. (2008) for the United Kingdom, Lawn (2008a) for Australia, Forgie *et al.* (2008) for New Zealand, Talberth *et al.* (2007) for the United States, and Castaneda (1999) for Chile.
3   GPI/capita was estimated by aggregating data for the 17 countries for which GPI or ISEW had been estimated, and adjusting for discrepancies caused by incomplete coverage by comparison with global GDP/capita data for all countries. All estimates are in 2005 US$.
4   This is most obviously shown with the 1973 OPEC-led oil embargo, which caused a sharp hike of the price of oil worldwide, and led to a world-wide recession. In the 1970s, partly led by the oil price surge, resource scarcities increased for all countries. Within the next nine years, most countries' GPI reached a threshold.
5   We need to recognize that Singapore has experienced considerable population growth during the period under consideration, and the value of many items, for example those that result from government investment such as *Services provided by infrastructure*, drop when population grows, and the benefits are estimated on a per capita basis, as we do here. Nevertheless, welfare is usually considered at a personal level, not at that of a country, and in many cases this lower welfare may be felt (for example when public transportation becomes more crowded, or roads more congested).
6   The Heritage Foundation and the *Wall Street Journal* publish an annual Index of Economic Freedom based on a number of indicators, including the effectiveness of the rule of law, regulatory efficiency, the openness of its market, and limited government intrusion. Hong Kong has ranked number one ('The freest economy in the world') every single year since the index was first created in 1994. The Singapore's economy has been rated less free, but it improved constantly, and in 2014 ranked second (Heritage Foundation, 2014).

References

Agrimoney. (2010). EU farm subsidies fall, bucking global trend. Retrieved 2 May 2013 from www.agrimoney.com/news/eu-farm-subsidies-fall-bucking-global-trend--1932. html

Bagstad, K. J. and Ceroni, M. (2007). Opportunities and challenges in applying the Genuine Progress Indicator/Index of Sustainable Economic Welfare at local scales. *International Journal of Environment, Workplace and Employment, 3*(2), 132–153.

Bleys, B. (2008). Proposed changes to the index of sustainable economic welfare: an application to Belgium. *Ecological Economics*, 64(4), 741–751.

Castaneda, B. E. (1999). An index of sustainable economic welfare (ISEW) for Chile. *Ecological Economics*, 28(2), 231–244.

Clarke, M. and Lawn, P. (2008). Is measuring genuine progress at the sub-national level useful? *Ecological Indicators*, 8, 573–581.

Clarke, M. and Shaw, J. (2008). Genuine progress in Thailand: a systems-analysis approach. In: Lawn, P. and Clarke, M. (Eds). Sustainable Welfare in the Asia-Pacific: Studies using the Genuine Progress Indicator (pp. 260–298). Cheltenham: Edward Elgar Publishing Ltd.

Deaton, A. (2008). Income, health, and well-being around the world: evidence from the Gallup World Poll. *The Journal of Economic Perspectives*, 22(2), 53–72.

Dong, B., Gong, J. and Zhao, X. (2012). FDI and environmental regulation: pollution haven or a race to the top? *Journal of Regulatory Economics*, *41*(2), 216–237.

Forgie, V., McDonald, G., Zhang, Y., Patterson, M. and Hardy, D. (2008). Calculating the New Zealand Genuine Progress Indicator. In: Lawn, P. and Clarke, M. (Eds). *Sustainable Welfare in the Asia-Pacific: Studies using the Genuine Progress* Indicator (pp. 126–152), Cheltenham: Edward Elgar Publishing Ltd.

Gil, S. and Sleszynski, J. (2003). An Index of Sustainable Econommic Welfare for Poland. *Sustainable Development*, *11*, 47–55.

Heritage Foundation. (2014). 2014 Index of Economic Freedom. The Heritage Foundation. Retrieved 10 May 2014 from www.heritage.org/index/ranking

Hong, V. X. N., Clarke, M. and Lawn, P. (2008). Genuine progress in Vietnam: the impact of Doi Moi reforms. In: Lawn, P. A. and Clarke, M. (Eds), *Sustainable Welfare in the Asia-Pacific: Studies Using the Genuine Progress Indicator* (pp. 299–330). Cheltenham: Edward Elgar Publishing.

Inglehart, R. (1997). *Modernization and postmodernization. Cultural, Political and Economic Change in 43 Societies*. Princeton: Princeton University Press.

Jackson, T., McBride, N., Abdallah, S. and Marks, N. (2008). Measuring Regional Progress: Regional Index of Sustainable Economic Well-being (R-ISEW) for All the English Regions.

Knox, P., Agnew, J. and McCarthy, L. (2014). *The Geography of the World Economy*. London: Routledge.

Kubiszewski, I., Costanza, R., Franco, C., Lawn, P., Talberth, J., Jackson, T. and Aylmer, C. (2013). Beyond GDP: Measuring and achieving global genuine progress. *Ecological Economics*, *93*, 57–68.

Lawn, P. (2008a). Genuine progress in Australia: time to rethink the growth objective. In: Lawn, P. and Clarke, M. (Eds). *Sustainable Welfare in the Asia-Pacific: Studies using the Genuine Progress Indicator* (pp. 91–125). Cheltenham: Edward Elgar Publishing Ltd.

Lawn, P. (2008b). Genuine progress in India: some further growth needed in the immediate future but population stabilization needed immediately. In: Lawn, P. and Clarke, M. (Eds), *Sustainable Welfare in the Asia-Pacific: Studies using the Genuine Progress Indicator* (pp. 191–227). Cheltenham: Edward Elgar Publishing Ltd.

Lawn, P. and Clarke, M. (2008). Genuine progress across the Asia-Pacific region: comparisons, trends, and policy implications. In: Lawn, P. and Clarke, M. (Eds). *Sustainable Welfare in the Asia-Pacific: Studies using the Genuine Progress Indicator* (pp. 333–361). Cheltenham: Edward Elgar Publishing Ltd.

Lawn, P. and Clarke, M. (2010). The end of economic growth? A contracting threshold hypothesis. *Ecological Economics*, *69*, 2213–2223.

Makino, M. (2008). Genuine progress in Japan and the need for an open economy GPI. In: Lawn, P. and Clarke, M. (Eds). *Sustainable Welfare in the Asia-Pacific: Studies using the Genuine Progress Indicator* (pp. 153–190). Cheltenham: Edward Elgar Publishing Ltd.

Max-Neef, M. (1995). Economic growth and quality of life: a threshold hypothesis. *Ecological Economics, 15*(2), 115–118.

Neumayer, E. (2001). Pollution havens: An analysis of policy options for dealing with an elusive phenomenon. *Journal of Environment and Development, 10*(2): 147–177.

Pao, H. T. and Tsai, C. M. (2011). Multivariate Granger causality between CO_2 emissions, energy consumption, FDI (foreign direct investment) and GDP (gross domestic product): Evidence from a panel of BRIC (Brazil, Russian Federation, India, and China) countries. *Energy, 36*(1), 685–693.

Rosenberg, D., Oegema, P. and Bovy, M. (1995). ISEW for the Netherlands: Preliminary Results and Some Proposals for Further Research. IMSA, Amsterdam.

Stockhammer, E., Hochreiter, H., Obermayr, B. and Steiner, K. (1997). The index of sustainable economic welfare (ISEW) as an alternative to GDP in measuring economic welfare. The results of the Austrian (revised) ISEW calculation 1955–1992. *Ecological Economics, 21*, 19–34.

Stymne, S. and Jackson, T. (2000). Intra-generational equity and sustainable welfare: a time series analysis for the UK and Sweden. *Ecological Economics, 33*, 219–236.

Talberth, J., Cobb, C. and Slattery, N. (2007). The Genuine Progress Indicator 2006. Retrieved from www.lanecc.edu/sites/default/files/sustainability/talberth_cobb_slattery.pdf

Wen, Z., Yang, Y. and Lawn, P. (2008). From GDP to the GPI: quantifying thirty-five years of development in China. In: Lawn, P. and Clarke, M. (Eds). *Sustainable Welfare in the Asia-Pacific: Studies using the Genuine Progress Indicator* (pp. 228–259). Cheltenham: Edward Elgar Publishing Ltd.

8   Towards a Steady State Economy

Introduction

In the previous chapters we have seen how the Singapore GPI has hardly grown from the early 1990s, while the Hong Kong GPI has continued growing, probably because it has been able to use its proximity to China to its advantage. In Chapter 7 we have also seen that most countries have experienced a drop in GPI per capita since the early 1980s, as the social and environmental costs associated with economic growth outstrip the economic benefits (Table 7.1). The dropping – or slowing – GPI calls for alternative economic policies to these of the neoclassical economic growth model, which states that more economic growth is always better than less. In this chapter we introduce the alternative model proposed by the Steady State Economy (SSE) school. We do not have sufficient space to give an exhaustive review of the arguments put forward within the SSE school, so we give a short introduction and summarize the main points. Our goal is not that of recommending specific reforms. How national economies should be reorganized, so as to attempt to increase welfare rather than economic output, should be the subject of a nationwide debate, as different options are available, and there is a need to fine-tune them according to the conditions of each particular locale. Our goal is that of raising some points that may contribute to that debate.

We start with a description of the fallacies of the neoclassical growth model, and discuss why endless growth is not only impossible, since the economy is a subsystem of a closed system (the earth's biosphere), but also undesirable and not useful. We then return to the ecological footprint (which we first introduced in Chapter 2) to argue that the world economy is already too large, and has been too large since the mid-1970s. Finally, we introduced some of the policy reforms that we believe are needed to transform the growth economy into a Steady State Economy. We draw in particular on the contribution to the SSE literature by Herman Daly, as we summarize the main concepts.

The economy as a closed system

The biosphere is a finite, non-growing and materially closed system. The inflow of radiant energy equals the outflow of heat. Some mass from outer space (meteorites) reaches the earth, but this is negligible. This means that the earth is neither shrinking, nor growing. On the other hand, the economy, is an 'open' system with a 'digestive tract' which takes matter and energy from the environment in low-entropy form (raw materials), and returns it to the environment in high-entropy form (waste). The economy depends on the biosphere, as a source of raw materials for the products it transforms, and as a sink of its waste. It is clear that since the economy takes from the biosphere whatever it transforms and returns to the biosphere the waste it generates; it is physically constrained by the size of the biosphere. The economy cannot grow forever. It can only grow until it reaches the size of the biosphere. Unfortunately, the ecological footprint (below) tells us that the world economy reached the size of the biosphere in the mid-1970s. Since then, it has continued to grow, and by 2010, 1.5 Earths were required to meet the demands humanity makes on nature. This is clearly unsustainable. Yet, we expect the economy to continue growing, and indeed our economic system is based on perpetual growth.

Unfortunately, economists do not consider that the economy can only grow within limits. This is likely due to the historic origins of neoclassical economic theory. Economic theory was not developed at a time when such enormous growth of people and the economy were conceivable. Neoclassical economic theory has its origin in the eighteenth and early nineteenth centuries, with such thinkers as Alfred Marshall (1842–1924), Vilfredo Pareto (1848–1923), and Marie-Esprit-Léon Walras (1834–1910) among others. At these times, the world population was closer to 1.6 billion (in 1900) and the economy was over 30 times smaller than nowadays (Maddison, 2007; Bourguignon and Morrisson, 2002). It is no surprise that economists did not concern themselves with the costs of economic growth, and did not consider the possibility of a limit to growth. Unfortunately, at the beginning of the twenty-first century, we have reached a level of development which threatens our future, and we must concern ourselves with the environmental and social costs of economic growth.

We argue that our growth economy needs to be replaced with a Steady State Economy (SSE), an economy designed not to grow: 'a Steady-State Economy is one whose throughput remains constant at a level that neither depletes the environment beyond its regenerative capacity, nor pollutes it beyond its absorptive capacity' (Daly, 1992: 163). Pursuing a SSE does not mean that the economy cannot develop. A Steady State Economy allows for qualitative development, but not quantitative growth. As the earth is continuously developing, for example allowing for the development of new organisms, but is not growing in terms of increasing in size, a Steady State Economy can continue developing, creating new products that are superior to the ones they replace, but the economy cannot grow in size. New products can be developed, and technologies can be invented, for example to improve the transformation of raw materials. However, the economy should no longer grow. Development means better, improved conditions. Growth means more of the same. The world cannot provide the metabolic throughput – as a source of raw material and a sink – of a growing economy forever. In practice, since we now live in a world which is severely skewed – with northern countries having a GDP many times higher than that of southern countries – we would need to see a shrinking of the economies of northern countries, to allow for a growth of the economies of southern countries.

Fallacies of the argument for further economic growth

Economists argue that additional growth would help us solve problems such as poverty, unemployment, and environmental degradation. This might be true in poor countries with low GDP, although environmental degradation is usually caused precisely by economic growth. However, when the economy is highly developed, the persistence of unemployment leads one to question the ability of further economic growth to address unemployment. It may as well be argued that what is needed to address unemployment is structural changes in the labour market rather than further growth. Similarly, poverty is not necessarily addressed by further economic growth, if the structure that causes that poverty (e.g. unequal access to education, weak unions) is not addressed. The fact that poverty and inequality keep increasing in countries whose economy grows, shows that the hidden hand of the market, which economists say will address poverty simply through economic growth, does not work. Economic growth by itself, without further policies and structural changes, will result neither in a drop in unemployment, nor a drop in poverty. This argument for economic growth is rather weak.

One argument put forward, for example by the World Bank (WB) and the International Monetary Fund (IMF), is that rich countries should continue to develop, as this will provide a market for exports from poor countries, which in turn will help poor countries develop. In many cases this argument is rather weak. Exports from poor countries are more often than not natural resources or labour-intensive manufactured goods. When natural resources are exploited, these are often extracted with low safety standards, and lead to great pollution. Furthermore, the purchase of these raw materials by rich countries increases their prices, which makes them too expensive for these poor countries that do not have them. This prevents their development, rather than encouraging it. Labour-intensive manufactured products may pay better than other options available to the people (such as farm work), but the salaries they receive are still very low. Also, the profits from the extraction of natural resources or from labour-intensive manufacturing often end up in rich countries, either because the corporations are headquartered – or owned by people living in – a rich country, or because local investors spend the profits on rich countries' luxury products. Under these conditions it is difficult for the country to develop. This can go a long way towards explaining why so few countries have been able to developed over the last 50 years. Apart from the Asian Tigers, most of the countries that were poor 50 years ago, are poor now.

What would be the impact on the poor countries and world trade if rich countries' economies stopped growing? Poor countries would see their export of raw materials dwindle, and the price of raw materials would drop. While this would reduce the incomes from export, it would also make them more affordable to the people in poor countries. Rich countries would spend more money on research and development of new technologies to improve the efficiency of production, since this would be the best way for companies to make money. New approaches should be developed to facilitate sharing these new technologies, for example the purchase of the technology by the state, for free distribution among poor countries.

Relative and absolute scarcity

Another important concept is that of relative versus absolute scarcity. David Ricardo (1772–1823) developed the idea of relative scarcity. Ricardo argued that as products become scarcer, there is an increase in the price of these products, and a gradual shift to other, more 'abundant' products, whose price is now comparatively lower. This means that it is very difficult to completely exhaust a resource, since as it becomes scarcer, the price increases and alternative products are chosen instead.

The other approach is that of absolute scarcity. This is often related to the writing of a contemporary of David Ricardo, Thomas Malthus (1766–1834), and is referred to as Malthusian scarcity. Malthus hypothesized that population grows exponentially (1, 2, 4, 8, 16, 32, etc.), while food production can only increase arithmetically (1, 2, 3, 4, 5, etc.). The result is that we will eventually reach a point when the world population cannot feed itself, famine occurs and a large number of people die, until an equilibrium is reached again. Such arguments have been repeated more recently, for example by the Club of Rome, which concluded that 'if the present growth trends in world population, industrialization, pollution, food production, and resource depletion continue unchanged, the limits to growth on this planet will be reached sometime within the next 100 years' (Meadows *et al.*, 1972).

Economists refute the approach of Thomas Malthus. The arguments of economists are twofold. First, they point out that Thomas Malthus' projection did not actually happen. Second, they point out that humanity has never actually run out of something that could not be substituted by something else, which also did the trick: we have discovered new resources, and better (more efficient) ways to extract and use them. In reality, little research has been done to refute Malthusian scarcity. It is simply dismissed as being historically inaccurate (and indeed it has been, until now). Indeed, the pessimistic projections made by the Club of Rome in 1972 did not come about. Yet, it is logical that if populations grow and per capita consumption increases, then absolute scarcity increases. It is fair to say that practically all renewable resources (trees, soil, water, etc.) in the world are being used unsustainably. Some resources may experience more relative scarcity than others (in the sense that oil may be more scarce than coal), but they all experience absolute scarcity (in the sense that ultimately they are all finite). Barnett and Morse's (1963, in Daly, 1985) assertion that 'Nature imposes particular scarcities, not an inescapable general scarcity' (p. 11) might have made sense in 1963, when they postulated it, but is obviously wrong in an economy that is well beyond a sustainable level, and that is rapidly expanding. We cannot shift from one product to another when the first experiences more relative scarcity, without the second also becoming more scarce. Some things may become 'relatively' more scarce than other things, but everything is becoming 'absolutely' more scarce.

If one resource becomes more scarce than another, its price increases. If we gradually shift from one good that is becoming 'relatively' more scarce to another, and by so doing the second also becomes 'relatively' more scarce, both prices increase. However, only the one whose price increases more would be thought as becoming more scarce. What happens if all resources become 'relatively more scarce'? Simply, inflation occurs, but the price signal does not tell us that the resources are more scarce, and we need to do something about it. Simply, we ask for higher salaries to compensate for the inflation, as if higher salaries would make these resources less scarce.

The argument that we have discovered new resources, and better ways to extract and use them, is also questionable. While it is true that we have discovered new resources, these only form a small part of the basket of goods we consume. Instead, more often than not we have developed better ways to extract old resources: the extraction of resources has expanded geographically and become planetary. Take fish for example: we are fishing in more and more remote areas because we have overfished the areas closer to the coasts. While in the 1980s much of the South Pacific and Indian Oceans were not overfished, only 15 years later there was no ocean left that was not overfished (Pauly *et al.*, 2005; Jackson *et al.*, 2001). Yet, we fail to realize that fish are becoming 'absolutely' scarcer because technological advances have allowed the limiting of price hikes. When prices do increase, they obviously do not increase at the same rate for all fish species. Thus, if the price of haddock increases more than the price of cod we may conclude that haddock is becoming scarcer, but they are all being overfished, and they are all becoming scarcer. The same is true for all other goods. For example, wherever we live, the beef we consume may come from Brazil, the iron from Australia, and the paper from the US. The fact that over the last decades we have been able to consume more resources does not mean that the amount of resources has increased. It means that we have gotten better at extracting gradually scarcer resources. In fact, most (virtually all) resources are becoming scarcer.

In some cases, as we improve the technology to extract increasingly scarce raw materials, those resources become cheaper. While economic theory tells us that the price should indicate the relative scarcity of a product, in reality the price no longer tells us how scarce the resource is. We have lost the signal that should allow us to ignore scarcities because 'the market will take care of it'. At the same time, we are slow and inefficient in introducing measures to replace that signal (e.g. fishing quotas) which leads to the general overconsumption of scarce raw materials.

All this means that we do not realize that these resources are becoming scarcer. Indeed, as technology advances, we extract scarcer resources at a lower cost. We are now extracting resources on a much more global scale. Most natural resources are overexploited, but we do not realize it. As a consequence, when we do run out of resources, it will be a sudden shock at planetary levels. Those who can afford it, will still be able to purchase these resources, but those who cannot afford it will suddenly be left without them. Since this is the normal way the economy works, it might not be recognized as a signal of the increasing scarcity of resources.

Absolute and relative needs

Another important distinction is that between absolute and relative needs. We can relate our discussion here to that of John Maynard Keynes (1883–1946), who described these two kinds of needs. Absolute needs are those necessary for survival, and include food, shelter and clothes. These are independent of the society in which we live, the expectations society puts on its members, and the situation of fellow human beings. Relative needs are those that we have to make ourselves feel equal or superior to others. While we strive to fill them, they are not necessary for survival. These may include branded clothes and mobile phones. While absolute needs are finite, relative needs are infinite.

Three things are important here. First, the economy does not need to try to satisfy relative needs because these cannot be satisfied. Relative needs are infinite. Second, relative needs have always existed. When the economy was smaller, and people spent a greater time fulfilling absolute needs than they do now, they still tried to produce goods to fulfil their relative needs. However, it is this spurious consumption of goods that makes the world economy unsustainable. If we only produced goods to fulfil absolute needs, the world would be able to sustain a much large population. Third, orthodox economics ignores the distinction between absolute needs and relative needs. Orthodox economics says that all needs are legitimate, equally worthy, and can and should be satisfied by economic growth. That relative needs can be satisfied is a logical fallacy, but this is ignored by orthodox economics. Growth is supposed to satisfy relative needs, but this can never happen.

Relative needs cannot be satisfied by growth, because the relative satisfaction of each individual is cancelled out by everybody else's satisfaction of their relative needs. Only being better off relative to others satisfies relative needs, but it is impossible for everybody to be better off relative to everybody else. Competition for this spurious consumption will never cease, but such competition greatly increases consumption and the toll on the environment. It is impossible for everybody to be at the same time better off than everybody else, so the existence of relative needs should not be encouraged. Yet, it is through advertisement, which tells us that our happiness is just another purchase away. The consequence of making the distinction between absolute and relative needs is that moving to a SSE would cost little in terms of foregone happiness, once absolute needs are satisfied.

How large can, and should, the economy be?

The questions we need to ask ourselves is: at what level can the world economy no longer grow, and we have to turn to a Steady State Economy? Those who have tried to answer this question have got it wrong, and have been criticized for even trying (e.g. the Club of Rome). This question is very difficult to answer, because we continue to extract and transform raw materials in more efficient ways. Instead of trying to identify the year by which we will run out of resources (as the Club of Rome did), which is exceedingly difficult, we can try to determine how many resources we consume compared to the ones that the biosphere (that finite, not growing, closed system) produces. That is, the maximum size of an economy that is sustainable, i.e. that guarantees the metabolic throughput. This is what the ecological footprint attempts to measure.

The ecological footprint is a measure of the human demand on the Earth's ecosystems.[1] It is an accounting system for ecological resources, which compares how much biological capacity we have, and how much we use. On the demand side, the ecological footprint estimates how much land and water area a human population uses to provide what it takes from nature. This includes the areas needed to produce the resources we consume, and the ecosystems needed to absorb the waste we emit, such as carbon dioxide. On the supply side, the ecological footprint estimates how much biologically productive area is available to provide these services. World average biocapacity per person was of 1.8 global hectares (gha) in 2008.[2] This corresponds to less than two football fields, from which every person in the world should obtain everything it needs (food, clothing, furniture, etc.) and dispose everything it uses. Each person inherits this 1.8 gha from its parents, and leaves it to its children. In addition, since there are also many other species in the world, we should leave some of these 1.8 gha untouched (the exact share should be a matter of debate). Put this way, it is obvious that the world biocapacity is very limited and needs proper management, something that is clearly missing now. Furthermore, the biological productive area is unfortunately shrinking, since many areas are gradually degraded or destroyed, and therefore cannot fulfil their functions as well as they did in the past. For example, as the amount of forest cover is reduced, the ability of forests to remove carbon from the atmosphere is diminished.

Since the industrial revolution in particular, the demand for resources has increased considerably. By the mid-1970s, humanity has been using more than the amount of biocapacity nature can regenerate (Figure 8.1). The Global Footprint Network calls this *ecological overshoot*. This means that since the mid-1970s the world economy has been unsustainable: we are using more resources than the environment can generate yearly, and we are disposing more than the environment can recycle yearly. We can do this, because over the past centuries we have been using less. Organic matter that was let to accumulate over the last millions of years can now be used in the form of fossil fuels. Forests which grew over the last centuries are now being cut. Glaciers which have existed for millennia are now melting, and we are using the water to grow food. We have started to

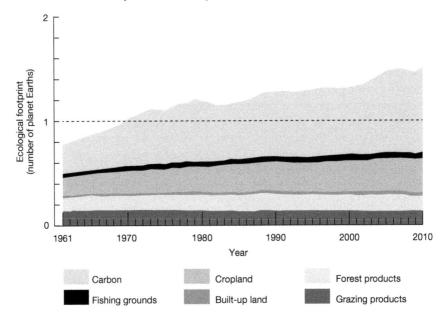

Figure 8.1 Humanity's Ecological Footprint, 1961–2010

Source: Global Footprint Network (2014b)

erode natural capital, that took centuries, millennia or millions of years to accumulate, and on which life depends. This is like using savings for an unsustainable lifestyle: since the mid-1970s we are behaving like an individual who uses his savings to increase his level of consumption. Surely this cannot last forever. Yet we are not addressing the problem and are behaving as if the ecological overshoot was of no concern. We may be able to continue with this unsustainable lifestyle for decades, but it is clear that one day we will run out of savings. The sooner we change our economy to one that respects the natural limits given by the earth's biocapacity – a steady state one – the easier the transition will be, since we will still have room for mistakes.

If we use the growth projections of the United Nations, by 2050 we will be using about twice the biocapacity of planet earth. Surely this is impossible, and even technological improvements will not be able to reduce this to a sustainable level. Currently if everybody lived as North Americans live, we would need five earths. For everyone to live as Europeans live, we would need three earths. The ecological footprint of Hong Kong was of 4.7 gha per capita in 2013 (Mattoon, 2013), and that of Singapore of 6.10 gha in 2012 (Global Footprint Network, 2012), similar to many rich countries but almost double the world average and around three times the 'sustainable level' of 1.8 gha. On the other hand, Hong Kong's biocapacity was of only 0.03 gha per person (Ewing *et al.*, 2010) and Singapore's biocapacity was of only 0.02 gha per person (Global Footprint Network, 2012). Exceeding local biocapacity is expected, since Hong Kong and

Singapore are small city-states, and is acceptable, but only as long as other areas do not use their whole biocapacity. Unfortunately there are fewer and fewer countries with additional biocapacity than the one they use.

Unfortunately, most people in countries with a lower ecological footprint aim to live the lifestyle of north American, European, Singaporean, or Hong Kong people. This is clearly not possible. In the name of fairness, it would be good if every person in the world could use the same amount of biocapacity (up to 1.8 gha). This means that people living in rich countries would need to make drastic cuts in their levels of consumption, to allow for greater consumption levels in poor countries. In line with the current thinking of creating markets to optimize the allocation of resources, and for example tradable emission permits for greenhouse gases, it could be conceived to organize a market whereby a country that overshot 'one world' would need to pay compensation to other countries to remain below 'one world' by the same margin. Unfortunately, the countries that are more powerful and less amenable to join international agreements that do not favour them, are also the ones that have a greater ecological overshoot, so this is unlikely to happen in the foreseeable future.

Mainstream economists, who are the most influential in setting government policies, do not ask themselves how large the economy can be relative to the ecosystem, and assume that the economy can grow forever. A keyword search using 'ecological footprint' in the ten academic journals in economics with the highest impact factor did not find a single article with that keyword. It is as if the topic was not relevant to economics, a discipline that studies the production, distribution, and consumption of goods and services. As if all these things did not require the environment.

Two sets of information are useful to determine the maximum size of a sustainable SSE. First, the economy cannot expand beyond the carrying capacity, as estimated by the ecological footprint. Second we are better off if we stop growing when the GPI is maximized. While the ecological footprint tells us the maximum we *can* aim for, beyond which our economies will collapse, the GPI tells us how much we *should* aim for, beyond which economic growth is no longer desirable. The size of the economy should be the lowest of the two. The ecological footprint should not be overshot, because that would make our economy unsustainable, and the GPI should not go beyond the threshold, because that would be undesirable from a welfare point of view (the costs exceed the benefits).

Humanity's ecological footprint overshot world biocapacity in about 1976 (Figure 8.1). On the other hand, 1978 was the year with the highest global GPI (Figure 7.3). The near overlap of the two years is interesting, although it might as well be simple coincidence, because fewer countries are included in the GPI figures than the ecological footprint figures. Also, the GPI figures most notably exclude the poorest countries, which need to develop further. This means that rich countries cannot simply shrink their economies to the 1978 level, but need to shrink it further, to allow some leeway to the world biocapacity to handle the additional requirements of the poorer countries' growing economies. However, it is clear that if the richer countries' economies shrank to the level of the mid-1970s, and were then managed more sustainably (which would also increase the world biocapacity, for example through reforestation), we would go a long way towards a more sustainable world, with greater human welfare.

The SSE does not equal to a life of want. After all, if those living in rich countries shrank their economies to the level of the mid-1970s, they would have greater, not lower, welfare than now. We can live sustainably while simultaneously increasing welfare. However, if we wait and further degrade our environment, we can envision a world whose biocapacity will have decreased so much, that we will have to severely curtail our standard of living. If we wait until scarcities become visible, we will need sudden, drastic, changes in our lifestyle to rectify much worse environmental conditions.

Technological improvements

Proponents of the idea that the economy can grow forever argue that technological development would make it easy to increase output while reducing inputs and waste. It is true that technology has been improving, and we now need fewer raw materials to produce the same (or better) products than we did in the past. However, putting all our hope in not-yet discovered technology is overly optimistic. Also, technological developments are usually not geared towards saving natural resources, or working less, but towards developing new products that may satisfy people's relative needs. For example, even though the development of electronics has been very rapid during the last few decades, technology that decreases the amount of inputs used in products has been by far outweighed by the increasing amount of products we consume. The result is that we use more raw materials to produce electronics now than we did in past decades.

However, even if we assume that technological improvement can halve our ecological footprint,[3] the economy of the rich countries would still have to shrink. In 2008, Singapore had an ecological footprint of 6.10 gha per capita, and Hong Kong 4.7 gha per capita, while the sustainable level (the world average biocapacity) is (up to) 1.8 gha per capita. Even if the ecological footprint of Hong Kong and Singapore was halved, consumption would still need to shrink. Clearly, putting trust in technological development alone is not satisfactory. Technological development can indeed help, but more emphasis needs to be put in channelling it to improve sustainability, rather than in developing new consumer products that satisfy people's relative needs.

Technological development is also used to replace workers with machinery, rather than to reduce working time. To continue with the same level of employment, it is implied that people will need to consume more. A constantly increasing consumption, in spite of a finite world, is indeed a requirement for continued full employment in a growth economy.

For technology to play an important role in increasing sustainability, the government must push for research and development in the right sectors. The right emphasis cannot simply come about by consumer demand. Thus, there must be a qualitative change in the kind of products being invented, not just a quantitative increase in the products being invented (this will also be touched upon in the next section). However, ultimately no technology can abolish absolute scarcity, and the fact that we live in a finite world. Technology may be able to delay the time we need to adopt a SSE, but cannot avoid the fact that we need a SSE.

Policy reforms

Our economies are designed to keep growing, and constant growth is a requirement for the economies to function. Failure to grow is considered a 'crisis'. A steady state economy is not a failed growth economy. A SSE is designed to not grow. As such, there need to be some small tweaking in the way the economy works, but the differences are only minor. In particular, there would be more state intervention. This does not mean that the SSE would be a centrally controlled economy. The market economy would still prevail and be tasked with allocating resources. However, there would be different rules, and in some places additional constraints. A little personal freedom would need to be sacrificed, but not much. On the other hand, if we wait for too long, until a severe environmental crisis causes the growth economy to collapse, we will need to make many more sacrifices, and sacrifice much more personal freedom.

A SSE should actually be very easy to achieve. Many of the necessary advances in technology, in terms of recycling, producing energy sustainably and using energy more efficiently, have already been made, but have simply not been implemented. However, there would also need to be some structural changes. In the following pages we review some of the changes that may be needed. These are just some of the options available, and experts should weigh in on – and fine-tune – the best approaches. We believe that a debate on the reforms necessary is urgently needed, and we hope that the following pages can contribute to that debate.

Products revolution

Very often natural resources are free (only their transformation is being paid for), energy is sold below cost (the high negative externalities are not included in the price), and the costs of disposing products is undervalued (the environmental costs of disposing or burning rubbish are not accounted for). This leads to waste of raw material and energy, and excessive consumption.

The SSE would introduce a cap on the amount of raw materials and energy used every year to a sustainable level. Within a national economy, producers would only be able to use a given amount of resources. The only way to increase the level of consumption with a SSE would be to increase the quality of the products. Instead of producing more output, there would be pressures, from producers, consumers, and the state, to increase the quality of resources produced with any given amount of raw material or energy. Hence, research and development would be focused on producing more goods with the same amount of raw materials, or producing goods that last longer and which consume less during their lifetime, and which can be recycled.

To facilitate the transition to the consumption of better products, one approach could be that of introducing personal consumption vouchers. If the ecological footprints of all products can be estimated, and it certainly can with some degree of precision, it would be easy to distribute vouchers to let people consume products only until a particular amount of ecological footprint (e.g. 1.8 gha) are 'consumed' yearly. Vouchers could then be bought or sold depending on a person's chosen lifestyle. A similar idea is being discussed in terms of personal carbon trading in the UK (Fawcett and Parag, 2010). Another approach could be that of encouraging leasing equipment (from cars to carpets), with the lessor being responsible for the recycling of the product once the lease ends. Finally, governments can promote new technology to make it more readily available to others. For example, the duration of patents could be shortened, or governments could buy patents, at a price that would still guarantee a profit for those who invented the products, and distribute it for free or sell it for cheap, especially to poor countries that still need economic growth.

Control trade

While poor countries would be allowed to develop, they would be allowed to do so only insofar as their ecological footprint would correspond to their allocated biocapacity. Without any further condition, we can envision that the more polluting companies would locate themselves to poor countries to produce goods to be exported to rich countries. We could also envision that as the poor countries' ecological footprint reaches their allocated biocapacity, and they can no longer add new industries, we can have a condition whereby they would contain many polluting industries producing export goods, and the people would still be poor and have low welfare. Hence, there would need to be some controls over the trade between nations. One way to address this problem could be to account for the ecological footprint of production in these countries where the products are exported to. For example, the ecological footprint of mobile phones produced by a company in China would be accounted for in the ecological footprint of the country where these mobile phones are exported to. There is sufficient information now to estimate the ecological footprint of products, but more research would improve the accuracy of such policies.

Since the ecological footprint accounting should be done at a country level, and different industries have markedly different ecological footprints (e.g. finance versus heavy industry), even trade among SSE countries should be regulated in the same way. Otherwise countries with a large heavy industry sector might be able to grow less than countries with a larger financial sector.

Reduce poverty by redistributing income

SSE is not equal to a failed growth economy. SSE is an economy designed to not grow. As such, it has to address some of the things that are supposed to come about as a consequence of economic growth. One crucial (supposed) impact of economic growth is that of reducing poverty. The argument put forward by those professing economic growth is that of *the hidden hand of the market*: some of the benefit of growth will trickle down to the poor, even without any form of state intervention. This is used as an argument to deny the minimum wage, or redistribution through taxation (taxation is supposed to slow economic growth, which weakens the trickle-down effect of economic activities). Growth is used as a substitute for redistribution. 'Growth offers the prospect of prosperity for all, with sacrifices by none' Daly (2014: 14): poverty and injustice will be washed away by growth. This idea is actually rather absurd since many needs are relative needs, but it is one that many economists treasure. The idea is also disproven by historic records. We have discussed the high (and increasing) inequality in Singapore and particularly Hong Kong. The same is true for other countries. For example, in 1982 in the US the highest-earning 1 per cent of families received 10.8 per cent of all pre-tax income, while the bottom 90 per cent received 64.7 per cent. In 2012, the top 1 per cent received 22.5 per cent of pre-tax income, while the bottom 90 per cent's share had fallen to 49.6 per cent (Saez, 2013).

Since the Steady State Economy will not grow, redistribution of income through taxation will be necessary, and demanded by the masses. It may be argued that the best way to address poverty is to limit the range between the minimum incomes and the maximum incomes. What is the best range of salaries, one which encourages hard work while preventing an inefficient life of luxury by the 'leisure class' (Veblen, 1899/2007)? Plato considered it to be a factor of four. Universities and the civil service seem to manage with a factor of ten or less. However, the corporate sector has much larger, and growing differences. In 2007, the CEOs of America's 500 biggest companies received US$10.5 million in compensation, or 344 times the pay of the average American worker. The gap between CEOs and the minimum wage workers is even greater: in 2007, CEOs total compensation averaged 866 times the income of minimum wage employees (Anderson *et al.*, 2008). Few would say that these differences are justifiable by the productivity of the individuals involved. A factor of 20 would probably be more than sufficient to reward productivity and hard work.

Income redistribution may be done better by limiting the difference between incomes among the highest earning and lowest earning in a country. There is now considerable discontent about the salaries of the top managers, and a few countries (such as Switzerland) have introduced measures (or are considering introducing measures) to give shareholders a larger say in the salaries of managers (Thomasson, 2013).

Stabilizing population

In a SSE, world population should no longer grow. At present, world population grows by approximately 1 per cent a year, and we are set to reach 10 billion before the world population will level off around 2060. Population growth is now considered a problem by many population scientists, while population decline is often also considered a problem, because of the dwindling number of people in working age. Stabilizing the population is now advanced as a good idea by many, although many religious authorities still oppose it. Only if the population is stabilized can output be stabilized without per capita loss. As to the means by which population can be stabilized, this needs to be debated.

China has engaged in a one-child policy (relaxed in rural areas and among minorities, and more recently also in some cities), which has been fairly successful in slowing population growth. On the other hand, Boulding (1964) proposed transferable birth licenses. Every person would be allowed to have a child. Couples who want more than two children would have to buy a voucher from a couple who wants fewer. The more people who want more than two children, the higher the price of the voucher. People in poorer countries often want more children, partly because no pensions exist to cater for the elderly, and a system of pensions should be introduced in these countries. The end result of such policy should be that no more children would be born than the replacement rate. If population shrinks, than more consumption per capita will be possible. From this perspective, a drop in population levels would be a good thing. Initially, stabilizing the population would require either increasing retirement ages, cutting retirement pensions, or higher taxes. This is already happening in some rich countries with very low birth rates.

How to deal with immigration is open to debate. Most immigration is from poor countries to rich countries. Poor countries should be allowed to develop, while rich countries should follow a SSE. It would make sense that immigration from poor countries would lead to an equal expansion of the recipient country, but a smaller expansion of the sending country. However, these are small issues that can easily be addressed.

Shorten work time

In rich countries we can envision leisure time to gradually increase, while working time drops as technology improves. Technical progress in production methods, accompanied by a fairly stable output (or in rich countries a shrinkage of the economic output), would result in shortening the working time instead of additional consumption, as is the case in the growth economy.

Nowadays youth and elderly unemployment are particularly high and difficult to tackle. Economic growth is said to help reduce that unemployment, although many countries have not been able to do so. Without growth it would be difficult to promote the possibility (or illusion) that reforms are not needed to reach full employment. The SSE would need considerable reforms in this area. With a SSE, it is unlikely that a large number of new jobs would be created. Hence, the introduction of a SSE should be accompanied by an effort to give jobs to unemployed people. Since technology does not improve uniformly across industries, we can envision some industries where leisure time gradually increases, and others where it does not. This would necessitate a shift in employment from some industries to others, to keep leisure time uniform across industries, since people would expect this. This would force some retraining, which would become a much more common feature than at present, and necessitate much more investment. There would obviously be some loss in efficiency when the workers are retrained, but the net result would still be a gradual reduction in the work time. The increasing leisure time would give the choice of whether to work as much as now on a weekly basis, but retire earlier, or work less on a weekly basis, and retire at the same age as now.

Stop advertisement

In the richest countries, much of the growth in the economic output is not due to needs that need to be fulfilled. To fulfil needs, probably less than 30 per cent of today's GDP would be sufficient. The rest are luxury items, some of which make our lives more comfortable, but which are of little actual use. The role of advertisement is mostly that of trying to convince us that we need such items. A SSE would aim at minimizing the amount of these products, so as to limit the negative impact on the environment. This would imply (and necessitate) a cultural shift from the pursuit of happiness through the consumption of consumer goods, to the pursuit of happiness through more free time, spiritual fulfilment, or other non-material means. Consumption does not need to drop to bare survival, and for many people in rich countries little change is actually needed. But if the growth economy is allowed to continue unabated, the environment will be so degraded, in terms of the depletion of natural resources and its ability to recycling waste, that consumption will indeed drop to bare survival.

Ecological tax reform

In many cases, environmental degradation is externalized. That is, the price products are sold for does not include the costs of the environmental degradation that takes place when the products are produced or disposed. This is in fact a subsidy to environmentally degrading products: the higher the degradation (the costs to the environment) the higher the subsidy. This leads to the excessive production of environmentally degrading products (e.g. electricity produced by coal power stations) and insufficient production of more environmentally friendly alternatives (e.g. electricity produced by wind turbines). The tax code should be reformed to take into account these externalities. For example, coal power plants should bear the full environmental costs of coal burning, which they should charge the consumers. The consumers would then have an incentive to either save energy, or shift to other less polluting forms of energy. An ecological tax reform would also force industries to use raw materials more efficiently. Ecological taxes can also be used to encourage recycling solid waste (Pichtel, 2010). In most countries, although land filling is paid by taxpayers, the amount paid through taxes is not dependent on the amount of waste discarded. This does not encourage the reduction of solid waste.

Reform national accounts

Ideally an economy should stop growing when marginal costs equals marginal benefits, but because of the system of national accounts we use, we do not know what the real marginal costs and the real marginal benefits are. Gross domestic product (GDP) accounts for the flow of marketed resources, but does not include the value of the raw materials that are not extracted. Excluded are also some of the costs of manufacturing, using, and discarding the products produced, for example the air, water and soil pollution. To fully account for the size of the economy, and better understand when marginal costs equal marginal benefits, a SSE would need a system of national accounting that includes that information.

Reform the banking sector

Presently banks impose interest rates on borrowing. This implies, and necessitates, that the businesses that borrow money make a profit and grow. In a SSE businesses would no longer grow (some obviously would, but that would no longer be the goal of businesses). Interest rates would likely be lower than now, however not zero. This is because capital would still be scarce, people would still have a positive time preference, and the value of total production would still increase, not because of additional quantity produced, but because the quality of the products would gradual improve. As is the case now, for example for electronics, better products would command a higher price, which can allow for the payment of interests.

Future growth expectations also sustain a very large finance industry that speculates on future prices of raw materials, among others. A SSE would likely not sustain such a large finance industry simply because future needs for raw materials would be smaller and more easily predictable, thus sustain fewer speculators. On the one hand, this would free intelligent people from the finance industry, and lead them to find employment that is more useful to society. On the other hand, since much of this speculation is driven by credit, it is likely that the amount of debt in the economy would shrink.

Control and centrally allocate scarce resources

Some resources are inherently scarce, because of physical constraints. This can be said to be the case, to some extent, of housing in Singapore and Hong Kong. In these cases people work harder to access the scarce resources, but since supply remains limited working harder simply results in higher prices, and an endless rat race as people compete with others for these scarce resources. In this case, it is more efficient for government to take over the resource, and control their supply and demand. This has been done with various degrees of success in other countries (Abdul-Aziz and Jahn Kassim, 2011; Field, 2014; Chiu, 2013). Considering the particular characteristics of Hong Kong and Singapore, where a sizeable proportion of households already live in public housing (Deng *et al.*, 2013), we believe the scheme could successfully be expanded to include a larger proportion of the population.

Conclusions

Throughout much of mankind's history, economic growth has been negligible. Only in the last 200 years, and in particular the last 50, has the economy grown considerably, as has the human population. This has resulted in an increasingly unsustainable use of natural resources. The economy is a subsystem of the earth, since it uses the earth's raw materials, and requires the earth to decompose the waste it generates. The closer the size of the economy is to the size of the earth, the more it will have to conform to the functioning of the earth. That functioning is one of a steady state, rather than one of growth. Since 1976, the size of the economy is greater than the size of the earth, in the sense that the economy uses more biological material every year than what the earth is able to produce or decompose during that year (Figure 8.1). The economy is able to do so because it uses biological resources which have accumulated over the last centuries, millennia, or millions of years (from forests to fossil fuels). It is like if we used savings to support an unsustainable lifestyle. At the same time, this lifestyle does not make us happy, since welfare (in the rich countries) has been dropping since the late 1970s (Figure 7.3). It is obvious that the economy has to shrink.

In this chapter we introduced the concept of Steady State Economy. We believe that this concept can help generate ideas to reform the current economic system. The SSE is an economy that does not grow. However, the SSE is not a failed growth economy, it is *designed* not to grow. Hence, the SSE requires reforms of the present growth-oriented economic system. We introduced some of the reforms we feel may be needed. A combination of different approaches should be used, including command and control approaches and market approaches. Some of these reforms are likely controversial for many people, and there needs to be a nationwide debate as to which approaches to follow, and how to fine-tune them. Having aggregate quotas is essential, since relative prices of products will not address their over-consumption. Quotas set quantitative limits to the amount of resources that can be extracted. However, the market mechanism should be retained, and it is likely that the best outcomes can be achieved by combining quota and market mechanisms. For example, there can be a quota on the amount of timber that may be logged, to be set considering ecological criteria (how fast timber can grow back). Market mechanisms can then be used to decide which timber can be extracted, from where, and to allocate the timber to different uses. The higher prices for resources would lead to investment in research and development for resource-saving technologies. Quotas on fossil fuels would also lead to more efficient uses (where alternatives are more difficult to find) and research in renewable energy.

The exact shape of these reforms should be a matter of public deliberation, as different societies might adopt different approaches. Countries are faced with a choice of options, and which option to pursue should be a matter of debate. We hope that this book may contribute to that debate. The policies discussed in this chapter are only some of the options, and experts should weigh in on the best approaches, once the goals are decided. However, economists and policy-makers need to wake up to the need for a transition to a SSE, or mistakes will not be discovered and better ideas will not be developed. The question is not whether the growth economy needs to be reformed. It does. Beyond a certain point, additional growth has more costs than benefits. Given the law of diminishing marginal utility and increasing marginal costs, this is not unexpected. It does in fact fit with economic theory. The question is what form the SSE should take. Unfortunately, even though more voices are raised that a transition to a SSE is both desirable and necessary, institutional changes are not on the agenda of any country as of 2015.

The economic profession plays a particularly important role here, partly because it is the profession which more than any other promotes the idea of an economy that can grow endlessly, partly because of its influence in setting government policies, and partly because of its ability to design and organize tools by which the SSE can function. Yet, there is very little interest among economists to challenge the dogma of an endlessly growing economy. Unfortunately, until a sizable proportion of the population agrees that a transition to an SSE is not only desirable, but also inevitable, there is no hope that the necessary reforms will be introduced, or even discussed and fine-tuned.

We hope that this book will contribute to the debate in Hong Kong and Singapore as to whether further economic growth is both desirable and necessary, whether a SSE would be a better alternative, and if so what reforms are needed to implement a SSE. While it is easy to imagine an SSE, it is more difficult to implement a transition from a growth economy to an SSE, because people resist change. For a more thorough discussion of the SSE and the failures of the growth economy, beyond those we have been able to introduce in this chapter, we refer our readers to the writings of Herman Daly (e.g. Daly, 1991, 1997; Daly and Cobb, 1989), Brian Czech (e.g. Czech, 2000, 2013), Tim Jackson (e.g. Jackson, 2009) and others (e.g. Dietz and O'Neill, 2013; Gilding, 2011; McKibben, 2007; Odum and Odum, 2001). Additional insights can also be obtained from the Degrowth literature, including Georgescu-Roegen (1971, 2011), Heinberg (2011), and many others.

Notes

1   The concept was developed by Mathis Wackernagel and William Rees in 1992. Mathis Wackernagel is now the director of the Global Footprint Network.

2   A global hectare is a common unit that encompasses the average productivity of all the biologically productive land and sea area in the world. Biologically productive areas include cropland, forest and fishing grounds, and do not include deserts, glaciers and the open ocean. Using global hectares allows for different types of land to be compared using a common denominator. Equivalence factors are used to convert physical hectares of different types of land, such as cropland and pasture, into the common unit of global hectares (Global Footprint Network, 2014a).

3   Halving the ecological footprint would obviously be very difficult. While doubling energy efficiency may be achieved, it might be difficult to double everything. For example, bottles of water would need half as much plastic, planes would need half as much oil to fly, etc.

References

Abdul-Aziz, A. R. and Jahn Kassim, P. S. (2011). Objectives, success and failure factors of housing public–private partnerships in Malaysia. *Habitat International*, *35*(1), 150–157.

Anderson, S., Cavanagh, J., Collins, C., Pizzigati, S. and Lapham, M. (2008). Executive Excess 2008: How Average Taxpayers Subsidize Runaway Pay 15th Annual CEO Compensation Survey. IPS-DC.org faireconomy.org. Retrieved 14 September 2012 from http://legacy.plansponsor.com/uploadfiles/excessexecpayFAS20080825.pdf

Boulding, K. E. (1964). *The Meaning of the Twentieth Century*. New York: Harper & Row.

Bourguignon, F. and Morrisson, C. (2002). Inequality among world citizens: 1820–1992. *American Economic Review*, 727–744.

Chiu, R. L. (2013). The Transferability of Public Housing Policy Within Asia: Reflections from the Hong Kong-Mainland China Case Study. In *The Future of Public Housing* (pp. 3–12). Springer Berlin Heidelberg.

Czech, B. (2000). *Shoveling Fuel for a Runaway Train: Errant Economists, Shameful Spenders, and a Plan to Stop Them All*. Berkeley, California: University of California Press.

Czech, B. (2013). *Supply Shock: Economic Growth at the Crossroads and the Steady State Solution*. Gabriola Island, Canada: New Society.

Daly, H. (1991). *Steady-State Economics*, 2nd edition. Washington, DC: Island Press.

Daly, H. (1997). *Beyond Growth: The Economics of Sustainable Development*. Boston, Massachusetts: Beacon Press.

Daly, H. (2014). *From Economic Growth to a Steady State Economy*. London: Edward Elgar.

Daly, H. E. (1985). The circular flow of exchange value and the linear throughput of matter-energy: a case of misplaced concreteness. *Review of Social Economy*, *43*(3), 279–297.

Daly, H. E. (1992). Steady-State Economics Concepts, Questions, Policies. *The Social Contract*, *13*(3), 163–170.

Daly, H. E. and Cobb, J. (1989). *For the common good: Redirecting the economy towards community, the environment, and a sustainable future*. Boston: Beacon Press.

Deng, Y., Sing, T. F. and Ren, C. (2013). The Story of Singapore's Public Housing: From a Nation of Home-Seekers to a Nation of Homeowners. In *The Future of Public Housing* (pp. 103–121). Heidelberg: Springer.

Dietz, R. and O'Neill, D. (2013). *Enough Is Enough: Building a Sustainable Economy in a World of Finite Resources*. San Francisco, California: Berrett-Koehler.

Ewing B., Moore, D., Goldfinger, S., Oursler, A., Reed, A. and Wackernagel, M. (2010). The Ecological Footprint Atlas 2010. Oakland: Global Footprint Network. Retrieved from www.footprintnetwork.org/images/uploads/Ecological_Footprint_Atlas_2010.pdf

Fawcett, T. and Parag, Y. (2010). An introduction to personal carbon trading. *Climate Policy*, *10*(4), 329–338.

Field, B. G. (2014). Public housing and the promotion of homeownership. *Innovative Housing Practices: Better Housing Through Innovative Technology and Financing*, 343.

Georgescu-Roegen, N. (1971). *The Entropy Law and the Economic Process*. Cambridge, Mass.: Harvard University Press.

Georgescu-Roegen, N. (2011). *From Bioeconomics to Degrowth: Georgescu-Roegen's' New Economics' in Eight Essays*. London: Routledge Studies in Ecological Economics.

Gilding, P. (2011). *The Great Disruption*. London: Bloomsbury Press.

Global Footprint Network. (2010). Ecological footprint atlas 2010. Oakland (USA): Global Footprint Network. Retrieved 14 August 2014 from www.footprintnetwork.org/images/uploads/Ecological_Footprint_Atlas_2010.pdf

Global Footprint Network. (2012). Living Planet Report, 2012: Biodiversity, Biocapacity and Better Choices. Oakland (USA): Global Footprint Network. Retrieved 14 October 2014 from www.footprintnetwork.org/images/article_uploads/Living_Planet_Report_2012_final.pdf

Global Footprint Network. (2014a). Frequently Asked Questions. Retrieved 14 October 2014 from www.footprintnetwork.org/en/index.php/GFN/page/frequently_asked_questions/

Global Footprint Network. (2014b). Living Planet Report 2014: Species and spaces, people and places. Oakland (USA): Global Footprint Network. Retrieved 14 October 2014 from http://ba04e385e36eeed47f9c-abbcd57a2a90674a4bcb7fab6c6198d0.r88.cf1.rackcdn.com/Living_Planet_Report_2014.pdf

Heinberg, R. (2011). *The End of Growth: Adapting to our new economic reality*. New Society Publishers.

Jackson, J. B., Kirby, M. X., Berger, W. H., Bjorndal, K. A., Botsford, L. W., Bourque, B. J. and Warner, R. R. (2001). Historical overfishing and the recent collapse of coastal ecosystems. *Science*, *293*(5530), 629–637.

Jackson, T. (2009). *Prosperity Without Growth: Economics for a finite planet*. London, UK: Earthscan.

Maddison, A. (2007). *The world economy volume 1: A millennial perspective volume 2: Historical statistics*. Academic Foundation.

Mattoon, S. (2013). Hong Kong Ecological Footprint report 2013. Hong Kong: WWF.

McKibben, B. (2007). *Deep Economy: The wealth of communities and the durable future*. New York: Henry Holt and Company.

Meadows, D. H., Goldsmith, E. I. and Meadow, P. (1972). *The Limits to Growth* (Vol. 381). London: Earth Island Limited.

Odum, H. T. and Odum, E. (2001). *A Prosperous Way Down*. University Press of Colorado.

Pauly, D., Alder, J. *et al.* (2005). Marine Fisheries Systems, Geveva: UNEP. Chapter 18. Retrieved 14 November 2014 from www.unep.org/maweb/documents/document.287.aspx.pdf

Pichtel, J. (2010). *Waste Management Practices: Municipal, hazardous, and industrial*. CRC Press.

Saez, E. (2013). Striking it Richer: The Evolution of Top Incomes in the United States (updated with 2012 preliminary estimates). Berkeley: University of California,

Department of Economics. Retrieved 14 October 2014 from http://elsa.berkeley.edu/~saez/saez-UStopincomes-2012.pdf

Thomasson, E. (2013). Swiss back executive pay curbs in referendum. Reuters. Retrieved 14 November 2014 from www.reuters.com/article/2013/03/03/us-swiss-regulation-pay-idUSBRE92204N20130303

Veblen, T. (2007). *The Theory of the Leisure Class*. Oxford: Oxford University Press. (Original work published 1899)

Index

'n' refers to end notes.

eBooks
from Taylor & Francis
Helping you to choose the right eBooks for your Library

Add to your library's digital collection today with Taylor & Francis eBooks. We have over 50,000 eBooks in the Humanities, Social Sciences, Behavioural Sciences, Built Environment and Law, from leading imprints, including Routledge, Focal Press and Psychology Press.

Choose from a range of subject packages or create your own!

Benefits for you
- Free MARC records
- COUNTER-compliant usage statistics
- Flexible purchase and pricing options
- All titles DRM-free.

Benefits for your user
- Off-site, anytime access via Athens or referring URL
- Print or copy pages or chapters
- Full content search
- Bookmark, highlight and annotate text
- Access to thousands of pages of quality research at the click of a button.

Free Trials Available
We offer free trials to qualifying academic, corporate and government customers.

eCollections
Choose from over 30 subject eCollections, including:

Archaeology	Language Learning
Architecture	Law
Asian Studies	Literature
Business & Management	Media & Communication
Classical Studies	Middle East Studies
Construction	Music
Creative & Media Arts	Philosophy
Criminology & Criminal Justice	Planning
Economics	Politics
Education	Psychology & Mental Health
Energy	Religion
Engineering	Security
English Language & Linguistics	Social Work
Environment & Sustainability	Sociology
Geography	Sport
Health Studies	Theatre & Performance
History	Tourism, Hospitality & Events

For more information, pricing enquiries or to order a free trial, please contact your local sales team:
www.tandfebooks.com/page/sales

www.tandfebooks.com

For Product Safety Concerns and Information please contact our EU
representative GPSR@taylorandfrancis.com
Taylor & Francis Verlag GmbH, Kaufingerstraße 24, 80331 München, Germany

www.ingramcontent.com/pod-product-compliance
Ingram Content Group UK Ltd.
Pitfield, Milton Keynes, MK11 3LW, UK
UKHW021612240425
457818UK00018B/509